Edmund Johnson Spitta, Charles Slater

An Atlas of Bacteriology

Containing 111 Original Photomicrographs with Explanatory Text

Edmund Johnson Spitta, Charles Slater

An Atlas of Bacteriology
Containing 111 Original Photomicrographs with Explanatory Text

ISBN/EAN: 9783337337599

Printed in Europe, USA, Canada, Australia, Japan

Cover: Foto ©berggeist007 / pixelio.de

More available books at **www.hansebooks.com**

AN
ATLAS OF BACTERIOLOGY

CONTAINING ONE HUNDRED AND ELEVEN
ORIGINAL PHOTOMICROGRAPHS
WITH EXPLANATORY TEXT·

BY

CHAS. SLATER, M.A., M.B., M.R.C.S.Eng., F.C.S.
Lecturer on Bacteriology, St. George's Hospital Medical School

AND

EDMUND J. SPITTA, L.R.C.P.Lond., M.R.C.S.Eng., F.R.A.S.
Formerly Demonstrator of Anatomy, St. George's Hospital Medical School

LONDON
THE SCIENTIFIC PRESS, LIMITED
28 SOUTHAMPTON STREET, STRAND
1898

PREFACE

BACTERIOLOGY is so recent a science that from its infancy it has been able to record by the aid of photography the forms and characters of the micro-organisms discovered, and the pathological changes produced by them in the tissues. At an early date Koch insisted on the value of a photographic record as a convincing proof of the reality and accuracy of the descriptions. For the most part, proof of this kind is no longer needed, but more and more has photography taken the place of the diagram or the drawing. It is in Bacteriology that illustration by photography is perhaps more satisfactory than in any other branch of pathology. Numerous are the excellent illustrations occurring in works on this subject, but these plates, even when not deprived of some of the excellence of the original negatives by the process of reproduction, are generally scattered and fragmentary records of particular appearances. The *Atlas der Bakterienkunde* by Fränkel and

Pfeiffer is the only work with which we are acquainted which systematically illustrates by a series of photographs the life history of a specific micro-organism. Excellent as this work is, the cost prevents it from being one of the ordinary working books possessed by the student. The present work has adopted the idea of the "Atlas" in that it provides a series of photographs of preparations of micro-organisms and their cultures, such as are usually met with, and made by, the student in an ordinary course of practical Bacteriology. It is hoped on the one hand, that it may be a laboratory handbook to direct the attention of the student to the points which he should observe in his own preparations, at the same time helping the teacher by providing a series of grouped illustrations; whilst on the other it is thought it may find a place in the library of the medical officer of health and other practitioners, as an atlas to which on certain occasions he may find it convenient to refer. It is in no sense a text-book of Bacteriology, and the letterpress merely serves to link together and emphasise the teachings of the photographs. The illustrations are taken from fairly typical cultures and preparations, and though the appearances vary so much with varying conditions, it is hoped that those presented may serve to show, when compared with the specimens actually in the hands of the student, the direction and extent of the variation.

Descriptions of the methods employed in the preparation of the specimens have been omitted, as the student will necessarily possess some recognised text-book or work on practical bacteriology in which descriptions of methods are more fully and appropriately treated than they could be in such a work as the present. A statement of the method used in the preparation of each specimen forms part, however, of the short description affixed to the illustrations.

On the other hand, a short description of the photographic methods and apparatus employed would seem to find a fitting place as an introduction.

The number of illustrations is of necessity limited, and while trying to associate the importance of the subject with the completeness of the series, yet in some instances, such as tuberculosis, it is felt that the subject, so far at least as the histology of tubercle is concerned, is so fully treated and illustrated in text-books that it is needless to multiply representations by adding sections of tubercular organs.

PREFACE

All the specimens, with one or two exceptions which are recorded, have been made in the laboratory of St. George's Hospital. The reproductions are, without exception, from original negatives.

To those gentlemen who have kindly allowed their preparations to be reproduced we wish to express our thanks. We are especially indebted to Mr. Harold Spitta for his assistance in the preparation of the original photographs and in other ways. For the great care and skill with which the original photographs are reproduced we have to thank Mr. A. Dent of Messrs. Dent and Co., Process Engravers, Clapham.

CHAS. SLATER,
EDMUND. J. SPITTA.

October 1898.

CONTENTS

	PAGE
PHOTOGRAPHIC INTRODUCTION	1
BACTERIOLOGICAL INTRODUCTION	10
BACILLUS ANTHRACIS	24
BACILLUS TUBERCULOSIS	32
BACILLUS SMEGMATIS	40
LEPROSY (BACILLUS LEPRÆ)	41
BACILLUS MALLEI (GLANDERS)	46
PYOGENIC ORGANISMS	49
STREPTOCOCCUS PYOGENES	50
STAPHYLOCOCCUS PYOGENES, AUREUS AND ALBUS	54
MICROCOCCUS GONORRHŒÆ	57
BACILLUS TYPHOSUS	59
BACILLUS COLI	64
DIPLOCOCCUS PNEUMONIÆ (FRÄNKEL)	66
BACILLUS PNEUMONIÆ (FRIEDLÄNDER)	69
BACILLUS DIPHTHERIÆ	71
SPIRILLUM CHOLERÆ	77
SPIRILLUM FINKLERI	
SPIRILLUM AVICIDUM (METCHNIKOVII)	84
SPIRILLUM TYROGENUM (DENEKE)	

CONTENTS

	PAGE
BACILLUS PESTIS BUBONICÆ	87
SPIRILLUM OBERMEIERI (RELAPSING FEVER)	89
BACILLUS TETANI	90
BACILLUS ŒDEMATIS MALIGNI	94
BACILLUS ANTHRACIS SYMPTOMATICI	98
ACTINOMYCOSIS	99
PLASMODIUM MALARIÆ	103

INDEX . . 109

LIST OF ILLUSTRATIONS

	FIG.
Micrococcus Melitensis (*Malta Fever*)	1
Large Micrococcus (*Air*)	2
Diplococcus in Sputum	3
Streptococcus	4
Micrococcus Tetragenus	5
Sarcina Flava	6
Proteus Hominis (*Bordone Uffreduzzi*)	7
Bacillus Anthracis	8
Spirillum Rubrum	9
Spirillum Rubrum	10
B. Typhi Murium	11
B. Mycoides	12
B. Figurans	13
B. Lepræ (*Bordone Uffreduzzi*)	14
B. Anthracis	15
B. Anthracis	16
B. Anthracis	17
B. Anthracis	18
B. Anthracis	19
B. Anthracis	20
B. Anthracis	21
B. Anthracis	22
B. Anthracis	23
B. Anthracis	24

LIST OF ILLUSTRATIONS

	FIG.
Bacillus Tuberculosis in Sputum	25
B. Tuberculosis in Sputum	26
B. Tuberculosis in Sputum	27
B. Tuberculosis	28
B. Tuberculosis	29
B. Tuberculosis in Urine	30
B. Tuberculosis Hominis	31, 32, 33
B. Tuberculosis	34
B. Smegmatis	35
Bacillus Lepræ (Bordone Uffreduzzi)	36
Bacillus Lepræ	37
Bacillus Lepræ	38
Bacillus Lepræ	39
Bacillus Mallei	40
Bacillus Mallei	41
Bacillus Mallei	42
Streptococcus Pyogenes	43
Streptococcus Pyogenes	44
Streptococcus Pyogenes	45
Staphylococcus Pyogenes Aureus	46
Staphylococcus Pyogenes Aureus	47
Staphylococcus Pyogenes Aureus	48
Micrococcus Gonorrhœa	49
B. Typhosus	50
B. Typhosus	51
B. Typhosus	52
B. Typhosus	53
B. Typhosus	54
B. Typhosus	55
B. Typhosus	56
B. Typhosus	57
B. Coli Communis	58

LIST OF ILLUSTRATIONS

	FIG.
B. Coli Communis .	59
B. Coli Communis .	60
B. Coli Communis .	61
Diplococcus Pneumoniæ (Fränkel)	62
Diplococcus Pneumoniæ . .	63
Diplococcus Pneumoniæ (Fränkel)	64
B. Pneumoniæ (Friedländer)	65
B. Pneumoniæ	66
B. Diphtheriæ .	67
B. Diphtheriæ .	68
B. Diphtheriæ .	69, 70
B. Diphtheriæ .	71
B. Diphtheriæ .	72
B. Diphtheriæ .	73
B. Diphtheriæ .	74
Spirillum Cholera Asiatica (Koch)	75
Sp. Cholera Asiatica	76
Sp. Cholera Asiatica	77
Sp. Cholera Asiatica	78
Sp. Cholera Asiatica	79
Sp. Cholera Asiatica	80
Sp. Cholera Asiatica	81
Sp. Cholera Asiatica	82
Sp. Finkleri	83
Sp. Finkleri	84
Sp. Finkleri	85
Sp. Avicidum (Metchnikovii)	86
Sp. Avicidum (Metchnikovii)	87
Sp. Avicidum (Metchnikovii)	88
Sp. Avicidum (Metchnikovii)	89
Sp. Tyrogenum (Deneke) .	90
Bacillus Pestis Bubonicæ .	91

LIST OF ILLUSTRATIONS

	FIG.
B. Pestis Bubonicæ	92
Sp. Obermeieri	93
Bacillus Tetani	94
B. Tetani	95
B. Œdematis Maligni	96
B. Œdematis Maligni	97
B. Œdematis Maligni	98
B. Œdematis Maligni	99
B. Œdematis Maligni	100
B. Anthracis Symptomatici	101
Actinomyces Hominis	102
Actinomyces Bovis	103
Actinomyces Bovis	104
Actinomyces Bovis	105
Actinomyces Hominis	106
Plasmodium Malariæ (Tertian)	107
Plasmodium Malariæ (Tertian)	108
Plasmodium Malariæ (Tertian)	109
Plasmodium Malariæ (Malignant Tertian)	110
Plasmodium Malariæ (Malignant Tertian)	111

Figures 5 and 75 are from preparations kindly lent by Dr. W. H. Dickinson.
Figures 7 and 36 are from preparations kindly lent by Prof. Bordone Uffreduzzi.
Figures 91 and 93 are from preparations kindly lent by Mr. E. L. Hunt.
Figure 100 is a preparation by Mr. Deller.
Figure 111 is from a preparation kindly lent by Dr. Manson.

PHOTOGRAPHIC INTRODUCTION

As there are manuals dealing with photomicrography in its various branches already in existence, a complete description of the methods and apparatus employed seems unnecessary, and it only remains to state which of the usual methods were selected, and what deviations from these methods were made to meet the requirements of particular cases.

A horizontal camera was employed throughout except in the few instances where the nature of the specimen compelled the use of a vertical apparatus.

The microscope stand invariably used was a Zeiss No. 1A. The arrangement for clearing the stage which this stand possesses was found very convenient when culture plates and similar objects were to be photographed.

Apochromatic objectives were always used: their superior definition and their inherent property of photographing equally well in all colours of the spectrum make them especially suitable. Most of the photographs of high magnification were taken with a Powell and Lealand apochromatic $\frac{1}{12}$ N.A. 1·43, specially made for the photography of bacteria. It possesses a very flat field without loss of central definition, but has an inconveniently short working distance. All the other apochromatics were made by Zeiss—viz., $\frac{1}{8}$ N.A. 1·4, $\frac{1}{6}$, $\frac{1}{2}$, 1 inch, a 35 mm. lens, and a Planar of 50 mm. focus.

Projection eyepieces were mostly employed, with the necessary camera extension to obtain the required amplification, but at times the ordinary compensating oculars for higher magnification were used in their stead. It is not a little curious that occasionally certain specimens were better rendered by the $\frac{1}{12}$ and a high eyepiece with short camera extension than with the projecting ocular and greater camera length.

With respect to Substage Condensers, a dry apochromatic (N.A. ·95) by Powell and Lealand was always used for powers over $\frac{1}{6}$, and an achromatic by Zeiss or one by Conrady for all the others, excepting for the 35 mm and Planar lenses, which performed best when a Nelson's quasi-achromatic doublet was employed.

A limelight mixed jet by Beard was the illuminant, and after several years' experience we have no hesitation in expressing great satisfaction with its performance.

Owing to the special colours of the stains employed in bacteriological specimens the use of "colour screens" is absolutely necessary, in order that sufficient contrast between the organism and the background may be obtained, and the resulting photograph be sharp, clean, and crisp. It is in the selection of this "colour screen" that the individuality of the photographer becomes apparent, but even after much experience it is not always possible to choose, without trial, the screen which will give the best result. In the selection of these light filters much stress has been laid by some authorities on the advisability of using such screens as give a pure monochromatic light. In the photography of bacteria this appears to be both practically and theoretically wrong. Where the object is to get the best possible photographic resolution of such objects as diatoms, with their so-called secondary markings, whose lines may be resolved into dots many thousands to the inch, then both theory and practice demand that light of the shortest wave length possible be employed. But where contrast alone is wanted, as in the photography of bacteria which have no ultimate structure comparable to the fine details of the diatom, it is a disadvantage to employ more colour than is absolutely necessary to secure the requisite contrast, as this merely increases the exposure without adding to the perfection of the final result. Screens of all colours were employed to secure contrast *only;* the one most commonly used was a screen of " pot-green " glass of medium density and $\frac{1}{8}$ inch thick. This appeared to be the most suitable for the majority of stains which we met with in the specimens.

Focussing in photomicrography is always a difficulty. Sometimes an ordinary ground glass was used, and at others an exceedingly fine variety made especially for us by Ross & Co. A Dallmeyer hand magnifier was used with this.

The disadvantage of using the finest ground glass is that the general lighting of the field is entirely lost, while, with the rougher variety, though the general field of view is well seen, the coarse grinding prevents accurate focussing without some difficulty, especially when the hand magnifier is used. The focussing screen in the present instance was so constructed that the glasses could be readily changed as thought best.

An even illumination of the field is important if

an unequal background in the print is to be avoided. The lime must be carefully moved from side to side and up and down to secure this result. A convenient stand for the jet, which is capable of being racked in both these directions with ease and uniformity, was made at our direction by Mason of Clapham.

The condenser was always centred for each objective, and critical light was used throughout.

If dark spots appeared on the picture the lime was turned on its axis, or the condenser just touched so as to remove the marks without affecting the definition.

Occasionally we have employed a second condenser between the light and the substage condenser. It serves to broaden out the illuminant, and is very useful to equalise the light over the field. Nelson's quasi-achromatic condenser we have found very useful for the purpose, fitted on a stand similar to that employed for adjusting the limelight previously described.

The *aperture* of the iris must not be contracted at all when photographing bacilli with high N.A., for, if so, white defraction marks will appear round each bacillus, and will spoil the final result. Only when using low magnification for tissues must it be closed for the purpose of preventing "flooding" of light; and even then, not usually more than one-third of the diameter of the back lens of the objective when looking down the tube of the microscope after removing the eyepiece. We have found that an additional iris diaphragm placed above the objective, known to some microscopists as a "Davis diaphragm," is of great service with quite low powers—it reduces the N.A. of the objective a trifle, it is true, but it frequently gives greater depth of focus and sharpens up the whole picture.

The negatives were all taken without exception on Edwards's Isochromatic medium $\frac{1}{4}$-plates and developed with hydrokinone and soda. Although the isochromatism of the emulsion is not all that could be desired, yet its uniformity and fine grain have always been so pronounced that we have never felt inclined to use plates by other makers, although we have tried most of them.

Exposure is always a difficult matter; but as a rough guide we may state that, using the pot-green glass to which we have previously referred, the isomedium plate, magnification at 1000 diameters, and a strong limelight mixed jet, the exposure varied from $\frac{3}{4}$ to $1\frac{1}{4}$ minutes, or less if the auxiliary lens was used. It is not easy without practice at all times to be positive whether the negative has been over or under-exposed, as the appearances presented do not conform as much as would be expected with those presented by over and under exposure in negatives of other subjects. Speaking again quite generally, the best guide we have found is to examine the background of the developed plate about whose exposure there is doubt, taking no notice of the appearance of the bacilli. If the background of the negative is of the right density for producing a grey ground in the print—a matter which experience will soon teach—then, if the proper contrast glass has been used, the bacteria should appear well and clearly defined and the print should be "plucky" and sharp, presuming, of course, the focussing has been correct. But if the background be faint in the negative the final picture will never be a good one, but usually flat and feeble in contrast. Here it is necessary to point out that the judgment of the photographer is not unfrequently put to the test as to the amount of contrast really required to make the bacilli stand out in the print. In some cases an almost white background is best, whilst at others a grey one is to be preferred. We have in cases been obliged to try the direct experiment, and choose that which appeared to give the better result. A faint background in the negative nearly always means under-exposure or extreme under-development, and no amount of subsequent intensification will cure it. Over-exposure appears as an intense blackening of the background as well as of the bacilli; the same effect is often produced by extreme over-development with a rightly-exposed plate.

It occasionally happens that a capital specimen in a bacteriological sense is not sufficiently stained for any coloured glass to produce the necessary contrast to make therewith a good photograph. Advantage then must be taken of the fact that the bacteria rarely possess any details of structure, so that the picture must be under-exposed for the bacteria, and then very freely developed with a strong and well-restrained developer. No amount of development will then bring out details of the bacteria, even if there were any, because of the under-exposure, but the prolonged development may serve to increase the contrast. In many cases this will produce results far above expectation.

The negatives were always developed with Thomas's hydrokinone and soda, until the back of the negative looked very decidedly grey; 1 gr. of bromide to the ounce of developer was added once or twice during the process. This seems to keep the bacilli clear and free from deposit. Should development have been carried too far, the negative may be cleared if it is worth keeping; so, too, if the light be not quite even, the thickened part may be thinned by using the same clearing solution with a brush. The cyanide and iodine reducer was usually employed for this purpose.

The tubes were all photographed with a 5-inch Ross-Goertz anastigmatic lens, the reflections off the sides of the glass being avoided by immersing the tube in a water bath. Limelight was thrown on the tube either from the front or back, occasionally employing a ground glass between the illuminant and the water bath.

For all purposes " backed " plates were used.

BACTERIOLOGICAL INTRODUCTION

THE theory that infectious diseases were due to the presence of living organisms and that the characters of such diseases could be referred to the nature of the "contagium vivum" is of considerable antiquity. It is only of recent years, however, that proof of this position has been obtained, and that a large number of infections in man and the lower animals have been traced to the action of parasitic micro-organisms which have gained access to the tissues. These micro-organisms are for the most part unicellular plants belonging to the lowest division of the vegetable kingdom. They are grouped under the name of Schizomycetes, or "fission fungi," as a sub-division of the Fungi; or, perhaps better, as Schizophytes (F. Cohn), and regarded as closely related to the Algæ.

Organisms belonging to other allied groups—*e.g.*, the Hyphomycetes and Blastomycetes are known to produce diseases of animals and plants, but nearly

BACTERIOLOGICAL INTRODUCTION 11

all the organisms represented in the following photographs are included amongst the Schizophytes.

The cells in this group are chiefly remarkable in that it has been impossible to demonstrate, with certainty, a differentiation between the protoplasm and the nucleus.

The Schizophytes fall roughly into three morphological groups :—

I. Cocci, comprising all those organisms whose cells are spherical.
II. Bacilli, comprising the rod-shaped or cylindrical-celled organisms.
III. Spirilla, comprising the spiral or corkscrew cells.

This grouping cannot be regarded as strictly scientific, but it is convenient. Owing to the undoubted polymorphism of these organisms a further convention is required to make it a practical classification. The bacilli are not unfrequently found in a coccoid form, or in such extremely short cylinders that it is impossible to distinguish them from cocci. If, however, the apparent cocci are known to assume the bacillary form under any conditions, then the organism is ranked under the group which may be regarded as representing its highest form, assuming the passage from cocci to spirilla as an ascent in the scale of plant life.

The method of multiplication common to the entire group of Schizophytes is that by fission, or cell division. By outgrowths from the cell walls advancing in two directions until they meet, the original cell is divided into two exactly similar daughter cells. The number of planes of division, the parallelism of these planes, and the association of the daughter cells resulting from the process of division give rise to various sub-groups of the three primary morphological divisions.

Cocci.—These are spherical cells showing no differentiation. The cells of different organisms of this group show very considerable variation in size, as may be seen in Figs. 1 and 2, the difference in the cells ranging from ·3 μ to 2-3 μ in diameter. In the same micro-organism considerable difference in the size of the individual cells may be observed.

The principal sub-groups of the Coccaceæ are—

(a) *Staphylococci.*—This name is given to cocci forming bunches or irregular groups of cells produced by the successive division of the spherical organisms; the planes of division are probably uniaxial, but the successive planes not parallel to one another.

Cover-glass preparations of organisms of this

FIG. 1.—Micrococcus Melitensis (Malta Fever).

Cover Glass Preparation, Agar Culture, Carbol-Fuchsine, apochromat, Projection ocular 6. × 1000.

FIG. 2.—Large Micrococcus (Air).

Cover Glass Preparation, Agar Culture, Carbol-Fuchsine, apochromat, Projection ocular 6. × 1000.

FIG. 1.—MICROCOCCUS MELITENSIS (MALTA FEVER).

Cover Glass Preparation, Agar Culture, Carbol-Fuchsine, $\frac{1}{12}$ apochromat, Projection ocular 6. × 1000.

FIG. 2.—LARGE MICROCOCCUS (AIR).

Cover Glass Preparation, Agar Culture, Carbol-Fuchsine, $\frac{1}{12}$ apochromat, Projection ocular 6. × 1000.

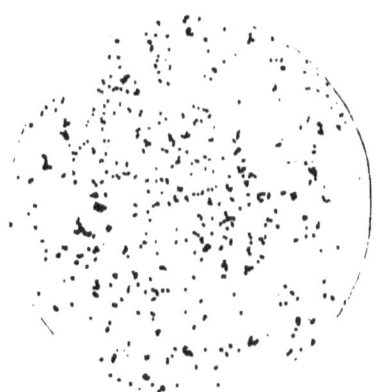

FIG. 1.

FIG. 2.

FIG. 3.

FIG. 4.

FIG. 3.—DIPLOCOCCUS IN SPUTUM.

Cover Glass Preparation, Gentian-Violet, $\frac{1}{12}$ apochromat. Projection ocular 6. × 1000.

FIG. 4.—STREPTOCOCCUS.

Cover Glass Preparation, Agar Culture, Carbol-Fuchsine. $\frac{1}{12}$ apochromat, Projection ocular 6. × 1000.

group are represented in Figs. 1 and 2. The cells are small, circular, and fairly regular in size in each specimen. The method of preparation prevents any conclusion as to the genetic relations of the groups. In Fig. 46, which represents the Staphylococcus Pyogenes Aureus in pus, the natural grouping of the organisms is better seen.

(b) *Diplococci*.—These are formed by the uniaxial division of the spherical cell and the coherence of the two resulting cells. On the further division of these cells the cohesion between the original pair apparently breaks down, so that successive pairs are formed. Fig. 3 shows diplococci occurring in sputum. The individual cells in this instance are somewhat elongated, and with a lens various stages of division may be observed, some of the longer cells being in a state of incipient sub-division. This lengthening of the cells which precedes division gives a bacillary form to the organisms. In some cases —*e.g.*, the gonococcus, Fig. 49, the division is not accompanied by elongation, so that the resulting diplococcus presents two flat opposed faces.

(c) *Streptococci* (Fig. 4).—These are organisms occurring in chains of varying length formed by successive uniaxial division of the spherical cell, the successive planes being parallel to one another, and the resulting cells remaining coherent. Obviously this group differs from the diplococci only in the coherence of the successive pairs of cells, and, indeed, it is not rare to find short chains of four or six members among the diplococci and pairs of cells among the streptococci. The not unfrequent dumb-bell-like arrangement in pairs of the cells comprising a streptococcus betrays this method of growth, and can be seen in several of the shorter chains in the photograph. The variation in size of the individual cells appears also.

The above are the chief sub-groups formed by uniaxial division of the spherical cell, but by a biaxial division in planes at right angles to one another the original cell gives rise to a group of four, and these tetrads are grouped under the name.

(d) *Merismopedia.*—The several varieties of the organism Tetragenus belong to this group, and that form which is frequently met with in cavities in phthisis is shown in Fig. 5. Micrococcus tetragenus is apparently harmless in man, but is pathogenic in the mouse, and from the peritoneal fluid of such an inoculated mouse the preparation is made. This slide shows, in addition, the fact that a micro-organism may be surrounded by a distinct capsule. The capsule is a very inconstant feature, being

FIG. 5. — *Micrococcus Tetragenus.*
Préparation, Splenic Fluid, Carbol-Fuchsin.
Projection ocular 6. × 1000.

FIG. 4. — *Spirosa Favre.*
Préparation, Gelatine Culture, Gentian-Violet.
Projection ocular 6. × 1000.

FIG. 5.—MICROCOCCUS TETRAGENUS.

Cover Glass Preparation, Splenic Fluid, Carbol-Fuchsine, $\frac{1}{12}$ apochromat, Projection ocular 6.   × 1000.

FIG. 6.—SARCINA FLAVA.

Cover Glass Preparation, Gelatine Culture, Gentian-Violet, $\frac{1}{12}$ apochromat, Projection ocular 6.   × 1000.

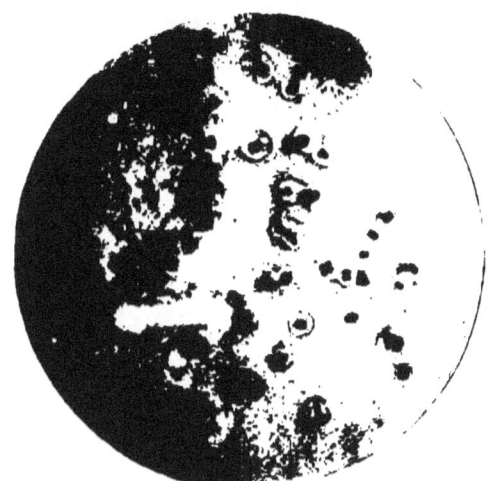

Fig. 5.

Fig. 6.

generally greatly dependent on the conditions under which the organism is living. It generally disappears in artificial cultures, and is best seen in the secretions or tissues of the infected animal. These remarks apply to most capsulated organisms.

It will be noticed that the capsule surrounds the group. Cocci collected into irregular groups surrounded by capsules are sometimes called *asco-cocci*. Beside the tetrad groups there appear in the preparation groups of two organisms. These probably represent the tetrad viewed from the side, or, in some cases possibly a group in which the division in the two axes has not been synchronous.

(e) *Sarcina.*—To this group belong those organisms which form groups of eight cells by a triaxial division of the original cell by planes at right angles to one another. The unit of division in this sub-group is a packet of eight. Such an organism is shown in Fig. 6. At first sight this would appear to belong to the preceding group, but this appearance is due to the impossibility of photographing organisms on two planes, and the consequent necessity for choosing a place in the specimen where the groups have been dissociated. The thickly-stained group on the left, and still better, the group on the extreme right of the specimen indicates the true character of the arrangement. The well-known *sarcina ventriculi* belongs to this group.

Bacilli.—The cylindrical organisms of this group differ from the Cocci in that division takes place in one direction only—at right angles to the long axis of the cell—so that no forms corresponding to Sarcina or Merismopedia occur. The bacilli vary very greatly in size, B. Anthracis, one of the largest of the pathogenic microbes, being from 3-5 μ long by 1 μ wide, while the microbe of influenza has dimensions approximately only the tenth of these. The breadth of the cell is a much more constant feature than the length, as the growth of the cell is in the direction of the long axis. As a rule the cylinder is of equal diameter throughout. Some possess much more rounded ends than others. By division and coherence groups of bacilli are formed which correspond to the diplococci and streptococci of the Coccaceæ, but these groupings are amongst the bacilli of less constancy, and the names which distinguish them are regarded more as describing varieties of a bacillus than as being the names of sub-groups. The term diplobacillus is not often employed, but might well be used for the organisms, which are by no means uncommon and were formerly described as figure of 8 bacilli (Pasteur). The chains

Fig. 7.

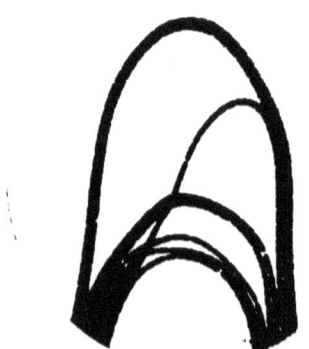

Fig. 8.

FIG. 7.—PROTEUS HOMINIS (BORDONE UFFREDUZZI).

Cover Glass Preparation of Culture, $\frac{1}{12}$ apochromatic, Projection ocular 6. × 500.

FIG. 8.—BACILLUS ANTHRACIS.

Edge of Impression Preparation of Colony on Gelatine, Löffler, $\frac{1}{12}$ apochromat, Projection ocular 6. × 1000.

of bacilli which correspond to the streptococci are called streptothrix filaments. Not unfrequently long threads are formed in which it is difficult to demonstrate the septa which divide the filaments into cells, and these threads are often spoken of as leptothrix filaments. The streptothrix filament is well seen in Fig. 8, together with the remarkable regularity with which the rows of filaments are arranged. The outgrowth of comparatively short bacilli into thread or leptothrix filaments is seen in photograph 7, which also shows many of the bacillary characters which have been mentioned, such as the constancy in diameter, the enormous variation in length, and the passage from forms which are practically cocci, through a succession of transition forms, to the long leptothrix filaments. Among the shorter forms are many diplobacilli or figure of 8 forms. The resemblance to a loosely-twisted spiral which some of the filaments exhibit is apparent only ; the threads are bent bacilli, and they do not form portions of a screw or spiral.

Spirilla.—To this group belong all organisms which occur as spiral cells or in forms derived from the spiral, and a glance through the illustrations of the various spirilla will show that there is great variety in the size of the various organisms, and in the length, regularity, and curvature of the cells.

The principal varieties of the spirilla which are named are—Comma bacilli, Vibrios, and Spirochætæ. The name comma bacillus was given by Koch to the Cholera spirillum; since it describes its usual shape, the name has remained in common use. The comma is really a segment representing a quarter turn of the spiral, and, like the spiral from which it is derived, varies greatly in the eccentricity of its curvature. Such comma forms constitute the great bulk of the organisms represented in Figs. 75, 83, and 86, and in Fig. 9.

The name *Vibrio* is often given to that derivative of the spiral which represents a half-turn such as is seen in Fig. 10; this may also be regarded as a double comma. Occasionally in these double comma forms, instead of an S-shaped organism, an E-shaped form arises from a reversal of the curvature.

The term spirochæta is applied to those spirilla, usually very long, in which the long axis of the spiral is itself curved and undulating.

Figs. 9 and 10 represent various forms of the same spirillum. The upper figure is from a growth on a solid medium, and it is in such media that the

FIG. 9.—SPIRILLUM RUBRUM.

Cover Glass Preparation, Agar Culture, Watery Fuchsine, $1\frac{1}{2}$ apochromat, Projection ocular 6. × 1000.

FIG. 10.—SPIRILLUM RUBRUM.

Cover Glass Preparation, Bouillon Culture, Carbol-Fuchsine. $1\frac{1}{2}$ apochromat, Projection ocular 6. × 1000.

Fig. 9.

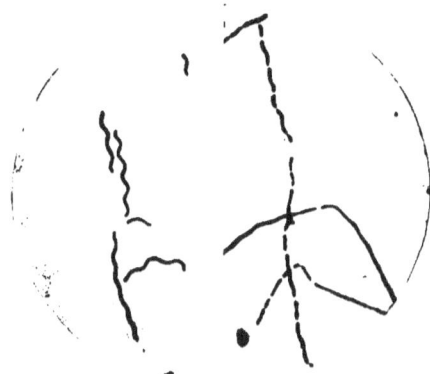

Fig. 10.

comma form is most usual, the spirilla form appearing only when the medium is so moist that the conditions resemble those of a fluid culture. In these latter (Fig. 10) the spirillum form is most common, and organisms of great length are found. In the right half of Fig. 10 the breaking down of the long spirillum with the formation of the derivative forms (comma bacilli and vibrios) is seen, and, in the lower portion, the remarkable flattening out of the spirals, which makes it difficult, not unfrequently, to distinguish these flattened spirals from bacilli.

Some of the cocci and a very large number of the bacilli and spirilla are motile. The activity of the movement varies greatly in different species, and in the same species under different circumstances. The bacilli appear to progress by rapid side-to-side or lashing movements; the spirilla, on the other hand, by a rotary movement round the long axis. These movements are due to the presence of flagella. The flagella are threadlike, wavy prolongations of the protoplasm of the cell, and are many times the length of the organisms to which they are attached. Their number and arrangement are very variable. Some organisms have a single flagellum situated at one pole of the cell (monotricha), as shown in Figs. 78 and 87; others a single thread at each pole (Amphitricha); while a third variety (Lophotricha) possesses at one extremity of the cell a bundle or brush of flagella. Perhaps the commonest arrangement is that shown in Fig. 11, in which flagella of varying number start from different parts of the periphery and surround more or less completely the cell body (Peritricha); these organs of motion can in some of the larger varieties be seen in the unstained condition, but the great majority require for their demonstration special staining methods.

Multiplication by fission is the method of reproduction common to the entire group of Schizophytes, but certain organisms are also capable of reproduction by spore formation. Spores are of two kinds—*arthrospores* and *endospores*. In those organisms which form arthrospores certain of the vegetative cells—*e.g.*, one member of a streptococcus chain (Leuconostoc)—becomes modified and acquires powers of resistance to harmful chemical and physical agents which are not possessed by the unchanged vegetative cells, while they at the same time retain the power of reproduction when placed under suitable conditions. It is a little doubtful how far these modified cells can be looked upon as true spores, and whether they should not be regarded

12.–13. Mycoides,
ver, Glass Preparatic
pachsine and Methyl
lar 6. × 1000.

n ocular 6. × 1000

FIG. 11.—B. TYPHI MURIUM.

Cover Glass Preparation, Flagella stained Löffler's method, $\frac{1}{8}$ apochromat, Projection ocular 6. × 1000.

FIG. 12.—B. MYCOIDES.

Spore Formation, Cover Glass Preparation, Agar Culture, Double-stained Carbol-Fuchsine and Methylene Blue, $\frac{1}{12}$ apochromat, Projection ocular 6. × 1000.

Fig. 11.

Fig. 12.

BACTERIOLOGICAL INTRODUCTION 21

as the resistant cells which occur in nearly every culture.

The formation of the Endospores is much better known, and occurs in certain of the bacilli and spirilla. Bacilli with endospores are shown in Fig. 11 and in several other photographs (17, 94, and 97) illustrating special organisms. The spore appears as a refracting granule within the cell in the position subsequently occupied by the developed spore, or as a number of scattered granules which coalesce to form the spore. In any case, as development proceeds, the granule grows and becomes an oval or spherical thick-walled, highly refractile body, while at the same time, as a rule, the protoplasm of the cell grows more and more translucent, and appears to be absorbed in the formation of the spore. The spores may, in relation to the mother cells, be central or polar. In Fig. 12 and in Fig. 17 they are central and do not deform the cell, the short axis of the spore being less than the diameter of the cell. In Fig. 97 they are central, and in Fig. 94 they are polar, but in both cases cause by their size a modification in the mother cell, causing it in the first case to become fusiform, and in the second to assume a drumstick character. These modified spore-bearing cells are known as *clostridia*. One spore only is formed within a single cell, but each cell does not necessarily form a spore.

The spore stains with difficulty and retains the stain, when once tinged, with considerable tenacity, so that it is possible to stain the spores and the body of the cells with contrasting stains. Fig. 12 is prepared in this way, the spores being red and the cells blue. Owing to the contrast between the strongly stained spores and the pale cell body, the latter appears narrower than the spore, but this is not really the case, as a comparison with unstained specimens at once shows.

The differences in the morphological characters of the various Schizophytes are frequently slight and variable, and in order to distinguish between them, advantage is taken of variations in the mode of growth on various media. Some liquefy gelatine, some do not; some form radiating colonies, others regular and circular colonies; some are granular, others structureless; some pigmented, others colourless. Many illustrations of the so-called characteristic growths of organisms on various media are given throughout the book. To permanently record the characters of the surface colonies of micro-organisms, preparations known as "impression preparations" are made. Such a one is shown in

FIG. 13.

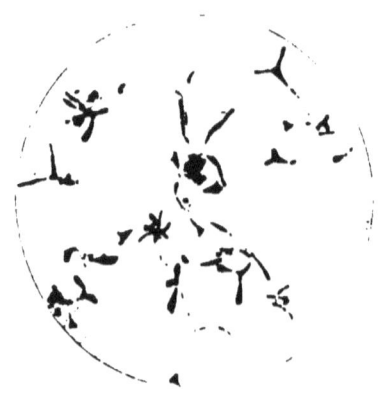

FIG. 14.

FIG. 13.—B. FIGURANS.

Impression Preparation, Gelatine Culture, Fuchsine, 24 mm. apochromat, ocular 5. × 60.

FIG. 14.—B. LEPRÆ (BORDONE UFFREDUZZI).

Cover Glass Preparation, Agar Culture, Carbol-Fuchsine, $\frac{1}{12}$ apochromat, Projection ocular 6. × 1000.

Fig. 13. They are made by pressing a cover glass on the surface of the growth and subsequently staining the adherent organisms. Under a low power (Fig. 13) the general character of the growth is well seen, while a similar preparation seen under a higher magnification (Fig. 8) shows the relation of the individual cells and of the filaments to one another, and displays the mode of growth which results in the production of the peculiar convoluted colonies with fibrillar offshoots which characterise the organism represented.

The mode of multiplication by fission which characterises the Schizophytes occasionally appears to be supplemented by budding or branching. Such budded or branched forms are shown in Fig. 14, which represents a sub-culture on agar of the B. Lepræ isolated by Bordone Uffreduzzi. Similar forms are seen in other bacilli and especially in B. Tuberculosis. How far these may represent forms of reproduction and how far they are merely forms of degeneration is still uncertain. They are certainly found most commonly in old cultures not remarkable for their vitality or virulence. The budding and branching have also been thought to betray a genetic relationship with the yeasts and hyphomycetes.

BACILLUS ANTHRACIS

In 1850 Davaine and Rayer observed that a non-motile bacillus was constantly to be found in the blood of animals which died of anthrax, and this observation was extended by Brauell in 1857 to the disease as it occurred in man. From 1863 onwards Davaine sought to prove that this bacillus was the cause of the disease, but it was Koch who, in 1876, furnished this proof, and it is to him that we owe our knowledge of the complete life cycle, including the spore-bearing stage, of the specific micro-organism. The study of the physiology of the bacillus, especially with regard to the possibility of obtaining cultures of attenuated virulence, and the use of such cultures as preventive vaccines, is in great part the work of Pasteur and his pupils. The appearance of the bacillus as it is met with in the blood of an animal dead of anthrax is shown in Fig. 15. It is a non-motile, rod-shaped organism, 3–5 μ in length and 1–1·5 μ in width, occurring

Fig. 13. B. Anthracis.

Smear Preparation, Splenic Blood (Cow, Carbuncular), Gram's method. Projection ocular 4. × 1000.

Fig. 14. B. Anthracis.

Glass Preparation, Splenic Blood of Mouse, Gram's method. Projection ocular 4. × 1000.

FIG. 15.—B. ANTHRACIS.

Cover Glass Preparation, Splenic Blood (Cow), Carbol-Fuchsine, $\frac{1}{12}$ apochromat, Projection ocular 6. × 1000.

FIG. 16.—B. ANTHRACIS.

Cover Glass Preparation, Splenic Blood of Mouse. Gram, $\frac{1}{12}$ apochromat, Projection ocular 6. × 1000.

Fig. 15.

Fig. 16.

in short chains of two to six members. The individual cells are square-ended, and stain strongly with any of the basic aniline dyes, and also by Gram's method. In stained specimens, owing to the original rectangular shape of the bacillus and the shrinkage due to the methods of preparation, the organism not unfrequently appears somewhat concave at the ends and sides, so that the chain of cells looks like a piece of jointed bamboo. The photograph shows some of the cells strongly and evenly stained, while others, owing to a degeneration or loss of their protoplasm, are irregularly or not at all stained, the position of the bacillus being indicated only by the faint outline of the cell envelope. This sheath, surrounding the protoplasmic contents of the cell, is occasionally very plainly seen, as in the preparation represented in Fig. 16. Though not always capable of being demonstrated, the sheath is a constant feature of the bacillus, and a development of the envelope has been regarded by Metchnikoff and Sawtchenko as, probably, a means by which the bacillus protects itself against bactericidal substances. In the blood, unless it has been shed for some time and kept under conditions favourable for the saprophytic growth of the bacillus, as occasionally happens in organs sent for examination, no trace of spore formation can be seen. The chains of bacilli also are short, unless the blood examined is that of a resistant animal inoculated with a virulent culture, or that of a susceptible animal inoculated with an attenuated organism; under these circumstances longer chains of bacilli may be found.

If a hanging drop bouillon culture be made from the blood, and kept at 37° C., the individual cells may be seen to divide and the resulting cells grow until they resemble the original mother cells. As the result of this division and growth the original single cells or short chains develop into enormously long, intertwisted, looped filaments, such as are seen in Fig. 8. When unhampered in growth, the chains of bacilli tend to arrange themselves with remarkable parallelism.

As the growth ages the division between the cells becomes more marked and the cells more cubical. At the same time, in the centres of many of the cells highly refractile points appear, gradually increase in size, and form an oval, highly refringent, thick-walled body—the spore—which lies with its long axis in the direction of the length of the cell and does not cause any alteration in the cell form. The protoplasm of the vegetative cell during this process becomes more and more hyaline until the spore comes to be free in the original cell envelope, and finally, by the disappearance of this envelope, is set entirely free. The preparation (Fig. 17) which is made from such a culture as that described shows the process of spore formation, which is unequally advanced in the various filaments. The spores are unstained and appear as clear oval spaces in the stained bacilli and are most advanced and best seen in the shorter filaments to the left of the photograph.

Spore formation requires the presence of free oxygen, and occurs at temperatures between 15° C. and 42° C. ; it is most free at about 20° C. to 25° C. The growth of the vegetative cell takes place at temperatures ranging between 12° C. and 45° C. ; the optimum temperature is about 37° C. The vegetative cell is killed at 60° C., and the spore at 120° C., though the mode of applying the heat and the length of time of exposure modify the result.

To the growth and formation of long intertwisted filaments and chains of bacilli, as described above, are due the characteristic microscopic appearances of anthrax cultures. In bouillon the growth forms cotton-wool-like masses, which remain coherent even when the culture is shaken. As the growth becomes older and spore formation advances, the cohesion between the cells grows less, and finally disappears when the spores are mature and free, so that shaking then produces a general turbidity of the fluid containing the growth.

The dull grey blanket-like growth of the young agar culture similarly betrays the microscopic structure.

In gelatine stab-culture (Fig. 18) the growth is very characteristic, and produces with its radiating filaments—starting from the line of inoculation, longest near the surface, and gradually diminishing as the depth in the gelatine increases, an appearance which has been compared to that of an inverted fir-tree. Liquefaction slowly takes place, commencing at the surface; and in the liquefied portion the growth acquires the usual cotton-wool appearance characteristic of young cultures in fluid media.

In gelatine plate cultures the appearance of the colonies differs as the growth occurs on the surface or in the depth of the medium. An impression preparation of a surface colony (Fig. 19) shows it to consist of a series of beautifully regular parallel lines of bacilli concentrically arranged, and resulting in a roughly circular growth with a waved margin.

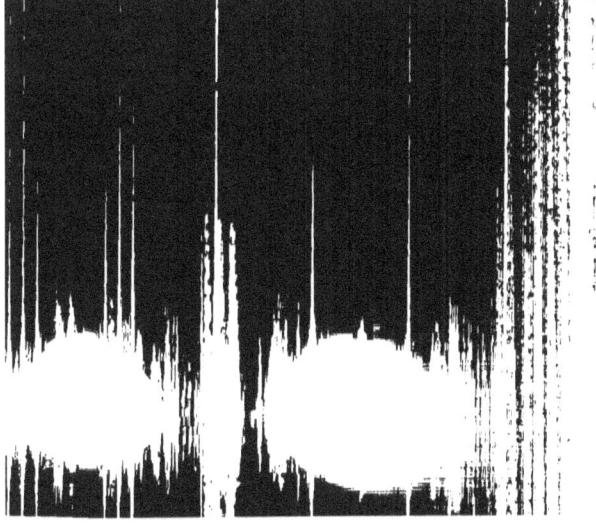

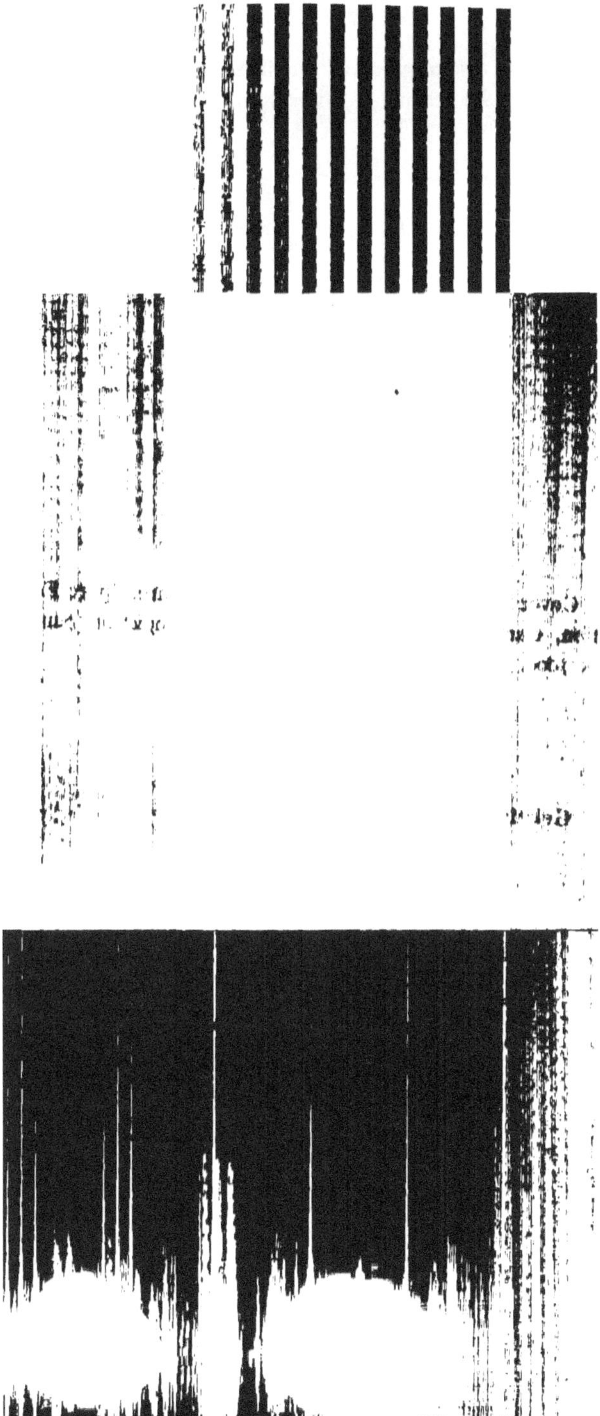

FIG. 17.—B. ANTHRACIS.

Cover Glass Preparation, Bouillon Culture, Spore Formation, Carbol-Fuchsine, $\frac{1}{12}$ apochromat, Projection ocular 6. × 1000.

FIG. 18.—B. ANTHRACIS.

Gelatin Stab Culture. 1 : 1.

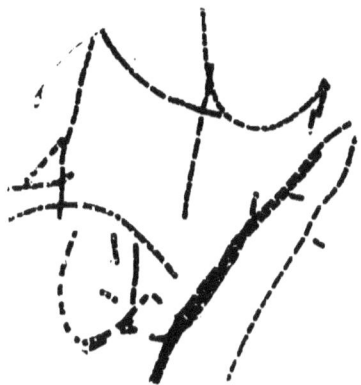

FIG. 17.

The intimate structure of the colony, and especially of the free margin, is shown in Fig. 8, which represents a highly magnified view of the edge of such an impression preparation.

In the colony represented the border is quite unbroken, but not rarely individual lines of bacilli break from the edge and pass over the surface of the medium, and occasionally the colony is surrounded by shaggy-hair-like processes (Medusahead colonies). In the depths of the medium the colonies are more spherical, granular, and brown. Slow liquefaction takes place.

Animals can be infected with anthrax either by inoculation or by the inhalation or ingestion of the bacilli. Though there are certain local differences due to the point of entrance of the bacillus, and though the symptomatology may differ, yet the microscopic appearances presented by the blood and by sections of the internal organs are in all cases those characteristic of an acute septicæmia. Beyond a general congestion there is very little structural change in the tissues, and the microscopic appearance presented is that of a general injection of the blood-vessels, especially of the smaller arteries and capillaries, with bacilli. Anthrax bacilli increase with enormous rapidity in the body of a susceptible animal, and are distributed in a passive manner, being themselves immobile, by the blood stream. The appearances presented in sections are conditioned by these circumstances. In man subcutaneous inoculation gives rise to a distinct local lesion, the "Malignant Pustule." This appears as a small vesicle seated on an indurated base. The centre rapidly becomes black and gangrenous, and the eschar is surrounded by other vesicles which unite with one another and with the primary lesion. There is great swelling and œdema of the tissues. A section through such a malignant pustule is shown in Fig. 20. The centre of the eschar is immediately to the right of that portion of the pustule which is shown. It will be seen that the papillæ are distended and disintegrated by the œdema, while they are at the same time crowded with bacilli, which also extend for some distance into the subcutaneous tissue. In the papillæ, as is best seen in the one at the right of the section, the bacilli are arranged in the direction of the blood-vessels.

In the lung, as represented in Fig. 21, the alveolar wall is but little modified, and the alveoli are free from inflammatory deposits, but the capillaries are filled by bacilli, and the alveoli are mapped out by this bacillary injection.

FIG. 19.—B. A. impression Preparation. Colo...
..let, Zeiss, 35 mm. apochromat.

FIG. 20.—B. A.
Section of Malignant Pustule,
ochromat, Compensating ocular

Fig. 19.—B. Anthracis.

Impression Preparation, Colony on Gelatine, Gentian-Violet, Zeiss, 35 mm. apochromat. × 20.

Fig. 20.—B. Anthracis.

Section of Malignant Pustule, Gram and Eosin, 24 mm. apochromat, Compensating ocular 12. × 125.

FIG. 19.

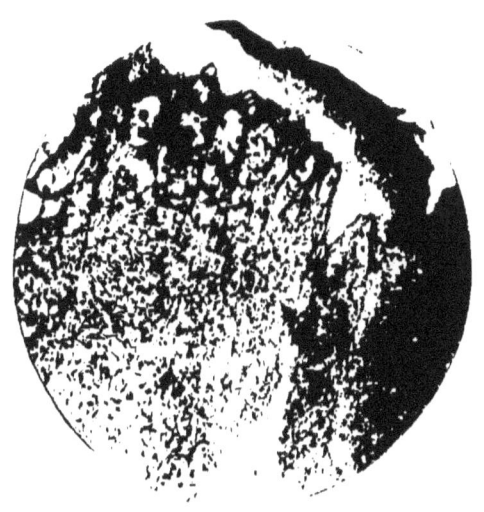

FIG. 20.

FIG. 21.—B. ANTHRACIS.

Section of Lung (Mouse), Gram and Eosin, $\frac{1}{A}$ apochromat, Projection ocular 6. × 400.

FIG. 22.—B. ANTHRACIS.

Section of Spleen, Gram and Picrocarmine, $\frac{1}{A}$ apochromat, Projection ocular 6. × 370.

FIG. 21.

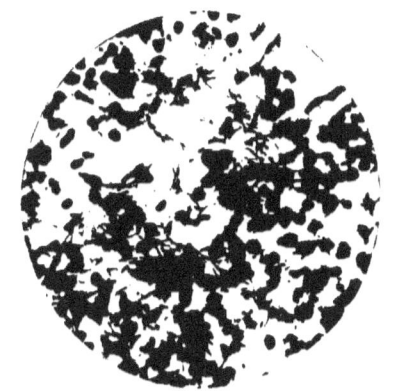

FIG. 22.

FIG. 23.

FIG. 24.

FIG. 23.—B. ANTHRACIS.

Section of Kidney (Glomerulus), Gram and Picrocarmine, $\frac{1}{8}$ apochromat, Projection ocular 6. × 375.

FIG. 24.—B. ANTHRACIS.

Section of Liver, Gram and Eosin, $\frac{1}{8}$ apochromat, Projection ocular 6. × 500.

In the spleen, which is always greatly enlarged in this disease, the bacilli are found in very great numbers (Fig. 22).

In the kidney the distribution of the bacillus is very characteristic, and is especially marked in the glomeruli and in their afferent vessels. These latter are often blocked with masses of bacilli, and the section (Fig. 23) shows the appearance of the glomerulus with its vessels filled with the microorganisms.

Similarly the section of the liver shows practically no histological change, but the bacilli are seen passing in lines between the columns of liver cells (Fig. 24).

BACILLUS TUBERCULOSIS

THAT tuberculosis was an inoculable and infectious disease was fully established by Villemin in 1868. This conclusion was confirmed and the paths of infection experimentally shown by the feeding experiments of Chaveau and Parrot (1868-1873), the inoculations of the anterior chamber of the eye, with the production of Tubercle of the Iris, by Cohnheim and Salomonsen (1877), and the inhalation experiments of Tappeiner (1878-1880). Numerous other workers also added to the evidence. The specific infectious agent existing in tubercular material was, however, not discovered until 1881-1882, when Koch demonstrated the constant occurrence of a minute, immobile, slender bacillus in the sputum of phthisical patients and in all forms of tubercular material.

The bacillus thus demonstrated was successfully cultivated. A repetition, with the pure culture, of the inoculation, ingestion, and inhalation experi-

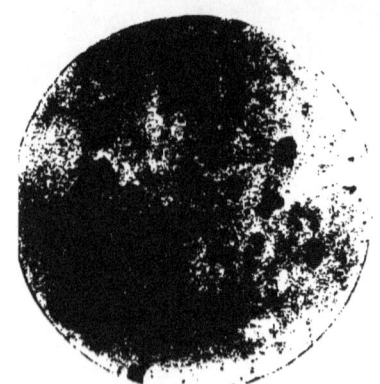

FIG. 25.

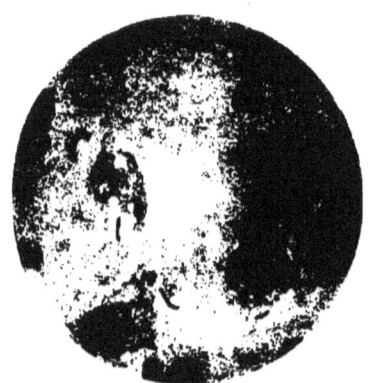

FIG. 26.

FIG. 25.—BACILLUS TUBERCULOSIS IN SPUTUM.

Cover Glass Preparation, Carbol-Fuchsine, Ehrlich method, counterstained with Methylene Blue, $\frac{1}{8}$ apochromat, Projection ocular 6. × 500.

FIG. 26.—B. TUBERCULOSIS IN SPUTUM.

Stained as Fig. 25, $\frac{1}{12}$ apochromat, Projection ocular 6. × 1000.

ments which had previously been made with impure material showed that the bacillus of Koch, contained in the inoculated matter, was the cause of the tubercular lesions produced, and confirmed the inoculability and infectiousness of the disease.

Subsequently Koch demonstrated in the cultures, or rather in extracts of the bacilli, a toxic material, "tuberculin," which has a specific inflammatory influence on tubercular tissues, and with which attempts have been made to cure or arrest the disease.

The bacillus (Figs. 25–29) is straight or slightly curved, usually round-ended, and varying in length from 1·3 to 3·5 μ—that is to say, about a quarter to half the diameter of a red blood cell, and is of extreme tenuity (·2-·4 μ). Originally demonstrated by staining with alkaline methylene-blue, it was almost at once shown by Ehrlich to have a so-called specific staining reaction, which displayed the bacillus clearly and differentiated it from all but two or three other micro-organisms (B. Lepræ, B. Smegmatis). This specific reaction depended upon the fact that, unlike nearly all other organisms, the tubercle bacillus, when strongly stained with fuchsine or gentian-violet, resisted the decolorising action of strong acids (*e.g.*, 25 per cent. nitric acid).

This power of resistance to the decolorising action of acids has been shown by Klebs and others to be due to the presence in the bacillus of a fatty body, extraction of which removes the specific staining property.

When stained by this method a specimen of tubercular sputum examined with a magnification of 500 diameters presents the appearance shown in Fig. 25. Amongst the scattered nuclei of cells are seen numbers of fine red-stained bacilli of very varying length and curvature, sometimes lying singly and sometimes collected into groups.

Examined with a higher power (Figs. 26 and 27), the bacillus usually presents a highly characteristic dotted or beaded appearance. Within a delicate sheath the protoplasm is collected into regular strongly stained masses separated by comparatively clear spaces. This appearance was originally ascribed to spore formation, but is now regarded as a segregation of the protoplasm without relation to spore formation.

This beading of the bacillus is by no means constant and may be entirely lacking as the photograps show. Not unfrequently there is a tendency for the bacilli to lie in pairs parallel to one another, and occasionally (Figs. 27 and 28)

FIG. 24. B. Tuberculosis in Sputum.

Cover Glass Preparation (Intracellular) stained as Fig. 25 ; apochromat. Projection ocular 6. × 1000.

FIG. 25. B. Tuberculosis.

Pus from Tubercular Cavity, stained as Fig. 25, 1/12 apochromat. Projection ocular 6. × 1000.

Fig. 27.—B. Tuberculosis in Sputum.

Cover Glass Preparation (Intracellular), stained as Fig. 25, $\frac{1}{12}$ apochromat, Projection ocular 6. × 1000.

Fig. 28.—B. Tuberculosis.

Pus from Tubercular Cavity, stained as Fig. 25. $\frac{1}{12}$ apochromat, Projection ocular 6. × 1000

FIG. 27.

FIG. 28.

FIG. 29.

FIG. 30.

FIG. 29.—B. TUBERCULOSIS.

Cover Glass Preparation, Culture on Glycerine-Glucose-Agar, Carbol-Fuchsine, decolourised by Sulphuric Acid 20 per cent., $\frac{1}{12}$ apochromat, Projection ocular 6. × 1000.

FIG. 30.—B. TUBERCULOSIS IN URINE.

Cover Glass Preparation, stained as Fig. 25, $\frac{1}{12}$ apochromat, Projection ocular 6. × 1000.

to form small groups of such regularly arranged bacilli.

In the sputum, as, indeed, in tubercular tissues, the bacillus may be entirely extra-cellular, or may occur singly, or in groups within the polynuclear pus cells. In Fig. 25 the great majority of the organisms lie without the cells whose nuclei are scattered over the field.

In Fig. 27, on the contrary, the bacilli are formed almost exclusively within the cells, many of which contain large groups of the micro-organism, and in some of these groups the parallel arrangement of the bacilli occurs.

It will be observed that in the specimens 25–28, two of which are preparations from sputum and one from the material found in a small tubercular cavity, there are practically no extraneous organisms. This, of course, is not always the case, and secondary infections play an important part in the processes of excavation ; but, speaking broadly, those portions of fresh sputum which contain many tubercle bacilli do not contain very many other microbes, though in neighbouring portions many varieties of organisms may be found.

When found in the urine (Fig. 30) from cases of tuberculosis of the urinary tract, the bacillus rarely presents the typical beading or bending, but usually stains in a homogeneous manner, and is found lying in clumps, occasionally of considerable size, and usually, as might be anticipated, associated with the pus cells contained in the urine. In all urine, but especially in the urine of women, the bacillus may be confused with the Smegma Bacillus (Fig. 35), which has a similar staining reaction, but is said to lose its stain when treated with alcohol sooner than the bacillus tuberculosis similarly treated.

From tubercles and tubercular materials, the specific bacillus was isolated and cultivated by Koch. Using gelatinised sterile blood serum as the cultivating medium, and spreading upon this surface the well-crushed material from an uncontaminated tubercle, faint traces of growth may be observed spreading from the inoculated material after about fourteen days' incubation at 37° C. The growth subsequently increases slowly, and forms small dry, scaly, raised, limpet-like crusts. The primary cultures are difficult to obtain, and not copious; but the sub-cultures are less exigent and can be grown with greater facility on other media, especially the agar medium containing glycerine 6 per cent. and sugar 2 per cent., which was originally suggested by Nocard and Roux. On this medium, unsuitable though it is for primary cultures, sub-cultures grow freely, forming a dry, wrinkled, or rather folded film (Fig. 31) of a slightly yellowish or, when old, pinkish tinge, and possessing a faint, sweet, mawkish smell. The growth may thicken (Fig. 32) and form roundish irregular corrugated heaps, and as in Fig. 33, this mode of growth may be so marked as to produce projecting, wrinkled, and folded "coral-island" like masses of considerable size, as may be seen in the photograph which represents the culture without magnification.

If the tube contains fluid, as it usually does, the growth tends to spread over the surface forming a film. This film growth is characteristic of the culture in fluid media, and the extent of the growth in solid media is largely dependent on the culture material remaining moist.

When freshly isolated the bacillus grows only within a narrow range of temperature; from 29° C. to 42° C. Its optimum temperature is about 37° C. After culture in artificial media for some generations the growth appears at an earlier date, and the culture is freer and more rapid than when first isolated, while the range of temperature within which growth occurs appears to extend.

When the glycerine-sugar-agar medium was first

introduced considerable stress was laid on this increase in saprophytic character, but, though some modification undoubtedly occurs, yet part of the effect attributed to the medium was due to the fact that *B. tuberculosis avium* was employed for cultivation, and this is now recognised as being a distinct variety differing considerably from the mammalian tubercle bacillus.

In culture (Fig. 29) the bacillus presents almost the same morphological characters and staining reactions as when in the sputum. The beading, though not so regular, is nearly equally marked, but the fine slender bacillus becomes, as a rule, thicker and coarser, and often presents enlarged extremities.

Under certain conditions long branched forms appear in the culture, which resemble closely the forms which are shown in Fig. 14. These forms have been studied by Coppen Jones, Babes, and others, and are usually an evidence of saprophytic life and small virulence.

Fig. 34 is a photograph of a section through a tubercular nodule stained with fuchsine and methylene-blue. The centre of the field is occupied by the characteristic giant cell, with the pale undifferentiated central portion of necrosed caseating

Figs. 31, 32, 33.—B. Tuberculosis Hominis.

Culture on Glycerine-Glucose-Agar. 1 : 1.

Fig. 34.—B. Tuberculosis.

Section of Bovine Lymphatic Gland, Giant Cell, stained as Fig. 25, ⅛ apochromat, Projection ocular 6. × 370.

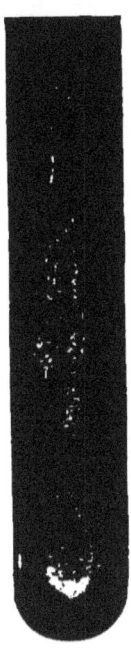  

FIG. 31.   FIG. 32.   FIG. 33.

FIG. 34.

material and well-stained peripheral fringe of nuclei. As to the origin of these giant cells and the histology of tubercle, it is not within the province of this book to speak. Within the giant cell and collected near the nuclei are seen numerous bacilli. The general arrangement is peripheral and radiate. In the tissue around the giant cell may be observed portions of the tubercle where necrotic changes are commencing, as shown by the poor and partial staining of the nuclei and other portions, where this change is so complete that the staining power of the tissue is completely absent.

BACILLUS SMEGMATIS

THE B. Smegmatis (Fig. 35) has already been referred to in speaking of the B. Tuberculosis in urine. It was originally described by Alvarez and Tavel, and its relation to the B. Syphilidis of Lustgarten suggested. The bacillus resembles closely in size and morphology the B. Tuberculosis, though it is somewhat more variable in length, and is not so definite in form; its extremities are often pointed. It is found in considerable numbers, and often in heaps, lying within and between the epithelial cells in the smegma. In its staining reaction it resembles the tubercle bacillus, but, though it resists decolorisation by acids and is thus differentiated from B. Syphilidis, its colour is removed by the action of alcohol. It has not been cultivated and is not inoculable.

FIG. 35.—B. SARCOMATIS.

From Gelatine Inoculation from Hepatic of Sarcoma, stained 25–35°; apochromat, Projection ocular 6, × 1000.

FIG. 36.—B. OF THE LEPRA (BORDONI-UFFREDUZZI).

Cover Glass Preparation of pure Culture, Carbol-Fuchsine stained with Nitric Acid 25 per cent., $\frac{1}{12}$ apochromat, Projection ocular 6, × 1000.

FIG. 35.—B. SMEGMATIS.

Cover Glass Preparation from Preputial Smegma, stained as Fig. 25, $\frac{1}{12}$ apochromat, Projection ocular 6. × 1000.

FIG. 36.—BACILLUS LEPRÆ (BORDONE UFFREDUZZI).

Cover Glass Preparation of pure Culture, Carbol-Fuchsine, decolourised with Nitric Acid 25 per cent., $\frac{1}{12}$ apochromat, Projection ocular 6. × 1000.

Fig. 35.

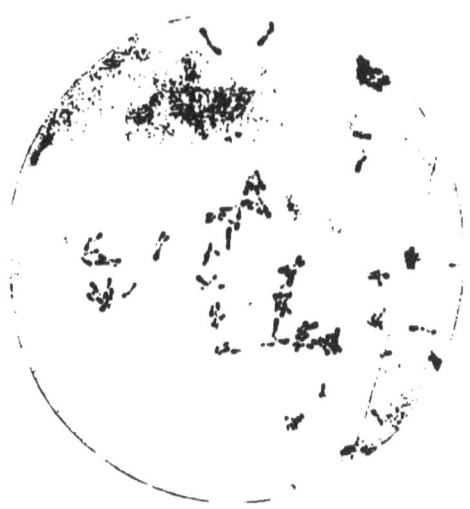

Fig. 36.

LEPROSY (BACILLUS LEPRÆ)

In the tissues of patients suffering from tubercular Leprosy Hansen, in 1874, discovered a bacillus which,.from its constant relation to the lesions and the large numbers in which it occurred in the diseased tissues, was, with probability, regarded as the cause of the disease. The same bacillus was subsequently discovered, but in very much smaller numbers, in the affected nerves and maculæ of cases of anæsthetic leprosy.

The bacillus found in the leprotic tissues closely resembles the tubercle bacillus, and is an immobile, fine, rod-shaped organism, 3–5 μ in length, reacting to stains as does the B. Tuberculosis, and showing the same beaded appearance. It is said to differ from the tubercle bacillus in that it is longer, straighter, stained more easily, and by methods which leave the tubercle bacillus unstained, and in that it resists decolorisation even more strongly than this latter bacillus. It also shows more irregularity in shape than the B. Tuberculosis, being sometimes pointed and sometimes swollen at the extremities. All these differential characters are inconstant and untrustworthy. Organisms corresponding to the typical form and variants from that form may be found in the photographs.

All attempts to procure undoubted cultures and successful inoculations with the supposed cultures have hitherto failed.

The most promising cultures have been obtained by Bordone Uffreduzzi (Turin) from bone marrow, and by Gianturco (Naples) from a non-ulcerated nodule. These independent observers appear to have isolated a bacillus which in morphology and staining reaction resembled the B. Lepræ of the tissues, and was not B. Tuberculosis. The bacillus isolated by Uffreduzzi is depicted in Fig. 36, and is characterised by the extreme segmentation and the swelling of the extremities. In late sub-cultures the specific staining reaction appears to be lost and the bacillus presents such branched forms as are seen in Fig. 14. These branched forms also occur in cultures of B. Tuberculosis.

Attempts to inoculate directly from the diseased tissues, whether in man or in animals, have not been much more successful than the attempts with cultures, though a measure of success seems to have been obtained by Arning, in man, and by Melchor and Orthman, in animals.

The bacillus has been found in nearly every organ, tissue, and secretion of lepers. Though occasionally, during the leprotic fever, found in the blood, it is chiefly spread by means of the lymphatics, and is associated with and produces its chief lesions in the connective tissues. The characteristic lesions produced are granulomata, but the true giant cell of tubercle is rare. In the skin (Fig. 37), after a preliminary stage of hypertrophy of the papillæ, these become flattened out and the tissue is the seat of a considerable sclerosis. In this sclerosed tissue islands of small-celled inflammatory growth occur, and in those nodules large numbers of bacilli are found both within the cells and in the intra-cellular spaces. Besides the great number of bacilli, the large "Lepra cells" constituting the circular, vacuoluted dark patches in the figure are highly characteristic of the leprotic lesions.

The origin of the lepra cells is still a subject of dispute, and opinions differ as to whether they are really cells crowded with bacilli or masses of bacilli occupying lymphatics. Though the former theory has not been disproved, there is a growing tendency, as a result of recent work, to regard these bodies as bacillary thrombi. On examination with a high power they are seen to consist of globular masses of bacilli crossing one another in all directions and lying embedded in a hyaline material which is possibly the result of the degeneration of the bacillus or its sheath. The bacilli tend to arrange themselves concentrically with the circumference of the mass, and to outline the vacuoles which occur in the cells (Fig. 39). The bacilli in these cells are usually very irregularly stained and much dotted, and are so intermingled that it becomes difficult to separate the individual microbes. This is the case with the two large masses shown in the photograph, but in the smaller and thinner group which lies midway between them, as also at the periphery of the larger masses, single greatly fragmented bacilli can be made out.

With a magnification of 1000 diameters a section of a leprosy nodule presents the appearance shown in Fig. 38, where the stiff, straight tubercle-like bacilli are seen scattered in very large numbers throughout the tissues, and lying some within the cells and some external to them. There are numerous transition forms between the small bacillus-containing cells of the granulation tissue

FIG. 37.—BACILLUS LEPRÆ.

Section of Skin, stained as Fig. 25, 24 mm. apochromat, Projection ocular 6. × 70.

FIG. 38.—BACILLUS LEPRÆ.

Section of Skin, stained as Fig. 25, $\tfrac{1}{12}$ apochromat, Projection ocular 6. × 1000.

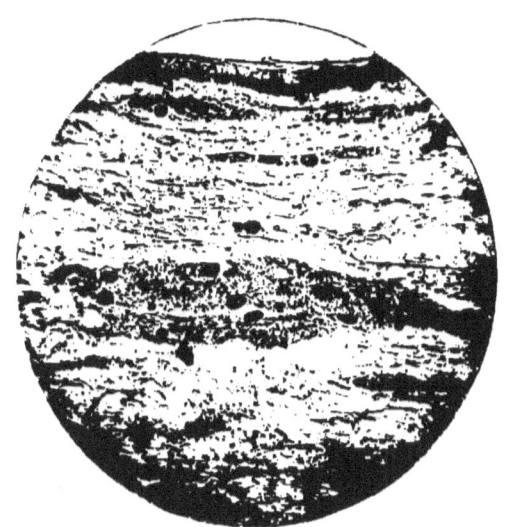

FIG. 37.

FIG. 38.

LEPROSY (BACILLUS LEPRÆ)

and the large cell to which, more particularly, the name Lepra cell is given. The latter are commonest in the older tissues. These bacilli do not extend beyond the Malpighian layer. The appearances presented by the skin are reproduced in affected mucous membranes, especially in the larynx. Affected nerves show an interstitial neuritis with large numbers of bacilli on the connective tissue. The internal organs, especially the liver, spleen, and testicle, are studded with granulomata crowded with the characteristic microbe.

That the disease is infectious to some extent there can be no doubt, but the mode of origin and the exact methods by which it is spread are still unknown.

BACILLUS MALLEI (GLANDERS)

THE lesions produced by this micro-organism are, like those occurring in Tubercle and Leprosy, ranged in the group of granulomata, and this constitutes a pathological link between the three diseases, though the B. Mallei differs considerably from the microorganisms of the other two disorders. The bacillus was isolated in 1882 by Loffler and Schütz, and the disease reproduced by the inoculation of the pure cultures.

The organism is found in large numbers in the rapidly suppurating lesions of glanders, and in the granulomatis, as well as in the pustules of the eruption which not unfrequently occurs in the course of the disease. A preparation from such a pustule is shown in Fig. 40. The bacillus, which is present in considerable numbers, is a small, straight, fine organism, somewhat resembling the B. Tuberculosis, but, as a rule, decidedly thicker. It measures usually $1\cdot5$-3 μ by $\cdot25$-$\cdot4$ μ. It is sometimes found

FIG. 39.—BACILLUS LEPRÆ.

Section of Skin—"Lepra Cell"—stained as Fig. 25, p. 79 throrunt, Compensating ocular 12. × 1000.

FIG. 40.—BACILLUS MALLEI.

Cover Glass Preparation, Material from Pustule Man, Carbol-Fuchsine, ¼, apochromat, Projection ocular × 1000.

FIG. 39.—BACILLUS LEPRÆ.

Section of Skin—"Lepra Cell"—stained as Fig. 25, ⅛ apochromat, Compensating ocular 12. × 1000.

FIG. 40.—BACILLUS MALLEI.

Cover Glass Preparation, Material from Pustule in Man, Carbol-Fuchsine, $\tfrac{1}{12}$ apochromat, Projection ocular 6. × 1000.

Fig. 39.

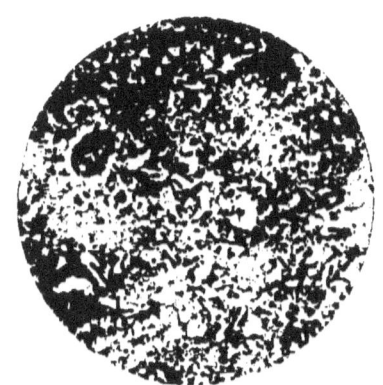

Fig. 40.

in pairs and occasionally, though rarely, in longer chains. It differs from the tubercle bacillus in the readiness with which it parts with its stain; all attempts at differentiation, and even the use of alcohol in dehydration robbing the bacillus of its colour. It is not stained by Gram's method. As in the tubercle bacillus, the protoplasm is segregated and the organism beaded, but very frequently, in the shorter forms at least, this segregation leads to polar staining, the bacillus presenting a clear unstained space in the middle while the two extremities are deeply coloured. Examples of " beading" and "polar" staining may be seen in the photograph.

Cultures from such purulent material can be obtained on agar (or, better, on glycerine agar), bouillon, blood serum, and potato at 37° C.

On agar it forms a rather thick and decidedly glutinous layer; in bouillon a general, rather slight turbidity, and on potato a thick, brown, viscid, honey-like growth, which is highly characteristic. Preparations from such cultures (Fig. 41) show bacilli of very varying length, ranging from coccoid forms to fairly long filaments. The cultures are very polymorphic and prone to the production of involution forms. The beading and polar staining occur in the cultures. The bacilli from cultures do not stain very readily, owing, it would seem, to the protective slimy envelope which gives to the growth the gelatinous character which has been mentioned. Susceptible animals—*e.g.*, guinea-pigs—succumb to subcutaneous and intra-peritoneal inoculations; the internal organs are studded with the characteristic lesions. In the male guinea-pig the testicles are peculiarly liable to suffer after intra-peritoneal injection, a severe epididymitis and orchitis being set up. This reaction in the animal is very useful in the diagnosis of glanders in man. A section of testicle from such an infected animal is represented in Fig. 42. The lesion figured is a very early one, and the alterations in the tissue correspondingly slight, but a group of bacilli is seen. The organisms are often met with in the cells.

A substance, "mallein," corresponding both in origin and in specific action to tuberculin, has been obtained.

It is doubtful whether susceptible animals have been successfully immunised.

FIG. 41.—BACILLUS MALLEI.

Cover Glass Preparation, Culture on Potato, 48 hours' growth, Carbol-Fuchsine, $\frac{1}{12}$ apochromat, Projection ocular 6. × 1000.

FIG. 42.—BACILLUS MALLEI.

Section of Testis of Inoculated Animal, Methylene Blue, $\frac{1}{12}$ apochromat, Projection ocular 6. × 1000.

FIG. 41.—BACILLUS MALLEI.

Cover Glass Preparation, Culture on Potato, 48 hours' growth, Carbol-Fuchsine, $\frac{1}{12}$ apochromat, Projection ocular 6. × 1000.

FIG. 42.—BACILLUS MALLEI.

Section of Testis of Inoculated Animal, Methylene Blue, $\frac{1}{12}$ apochromat, Projection ocular 6. × 1000.

Fig. 41.

Fig. 42.

PYOGENIC ORGANISMS

A VERY large number of organisms possess, amongst other properties, a pyogenic power; and there are not a few with whom this pyogenic function seems to be the principal one. Amongst these latter, however, there are three which, from the frequency with which they occur in localised collections of pus, in spreading purulent inflammation, and in pyæmia and septicæmia, are of much greater importance than the rest. These three microbes are: Streptococcus Pyogenes, Staphylococcus Pyogenes Aureus, and Staphylococcus Pyogenes Albus.

STREPTOCOCCUS PYOGENES

THE Streptococcus Pyogenes is associated with spreading inflammation of the erysipelatous type, but is also found in local abscesses and many other suppurative conditions. In pus (Fig. 43) it occurs in chains of spherical cells of extremely variable length, sometimes consisting of only three or four members, and sometimes of thirty or forty. The chains are usually free and extracellular. The individual cells are about 1 μ in diameter, are readily stained by any of the basic aniline dyes, and resist decolorisation when stained by Gram's method. Though frequently the organism consists of a series of regular and equidistant cells, yet often, as in the figure, the cells appear arranged in pairs separated by a smaller space than that which divides the pairs from one another. As has already been mentioned, in speaking of the varieties of micro-organisms, this is probably a result of the method of growth. In size of cells, irregularity in size of

FIG. 43.—Stereocomicus Process—
von Glass Preparation, Pus, Carbol Fuchsine, ¼;
etc.; Projection ocular 6. × 1000.

FIG. 44.—Streptococcus Pyogenes.
latine Culture. 1:1.

FIG. 43.—STREPTOCOCCUS PYOGENES.

Cover Glass Preparation, Pus, Carbol-Fuchsine, $\frac{1}{12}$ apochromat, Projection ocular 6. × 1000.

FIG. 44.—STREPTOCOCCUS PYOGENES.

Gelatine Culture. 1 : 1.

FIG. 43.

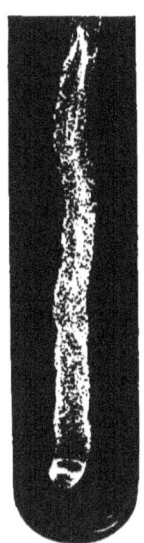

FIG. 44.

the individual cells, length of chains, and variability in the spaces between the cells the streptococcus shows much variety; many species have been described and named—*e.g.*, Strept. Longus, Brevis, &c. Streptococcus Erysipelatosus appears to be one of the varieties of this organism, and great differences in virulence are noted in cultures from various sources. From such pus as that described cultures of the organism—which grows between the temperatures of 18° C. and 42° C., with an optimum of 37° C.—can be obtained on gelatine or agar or in bouillon. The organism is a facultative anærobe. The growth on gelatine (Fig. 44) forms a series of small discrete, semitransparent, dewdrop-like circular colonies, and produces no liquefaction of the gelatine.

On agar the growth closely resembles that on gelatine, and there is often a very free growth in the fluid of condensation which collects at the bottom of the tube.

As in the morphological, so in the cultural characteristics, considerable variation is met with, affecting the freedom of the growth and the size and density of the colonies. In all cases, however, the colonies tend to keep separate and never reach any large size, while their vitality is not lasting, and unless frequently transplanted sub-cultures often fail. The media most suited to secure the continued vitality of the organism, and also to preserve its pathogenicity, are those consisting of admixtures of blood serum or ascitic fluid with bouillon (Marmorek, &c.).

Cover-glass preparations made from such cultures present the appearance shown in Fig. 45. The length of the chains of cells depends in part on the humidity of the medium ; the longer ones are found in bouillon cultures or in the condensation fluids of agar.

In fresh preparations from these fluids the organism is seen to form very long intertwisted and convoluted chains, which are generally much broken up in cover-glass preparations. It is probably due to this intertwisting that the microscopic characters of the cultures in bouillon are to be referred—viz., a primary turbidity with a subsequent rapid clearing of the fluid and deposit of flocculi at the bottom of the tube.

In cultures the same varieties are met with as have been described as occurring in pus, and are perhaps even more marked. The pathogenic affections produced by inoculations of the pure cultures vary greatly with the changing virulence of the

FIG. 45.—**Streptococcus Pyogenes.**

Above (Glass Preparation, **Bouillon** Culture, Gram.)
Zeiss D, Projection ocular C. × 1000.

FIG. 46.—**Staphylococcus Pyogenes Aureus.**

Cover Glass Preparation of Pus from Abscess, Gram
Eosin, A. apochromat. Projection ocular 6. × 1000.

FIG. 45.—STREPTOCOCCUS PYOGENES.

Cover Glass Preparation, Bouillon Culture, Gram, $\frac{1}{12}$" apochromat, Projection ocular 6. × 1000.

FIG. 46.—STAPHYLOCOCCUS PYOGENES AUREUS.

Cover Glass Preparation of Pus from Abscess, Gram and Eosin, $\frac{1}{12}$ apochromat, Projection ocular 6. × 1000.

Fig. 45.

Fig. 46.

organisms and with the method and seat of the inoculation. Subcutaneously inoculated, it may give rise to an erysipelatous inflammation, and the organisms are found filling the lymphatics at the margin of the inflamed patch; or it may give rise to a localised abscess or a spreading cellulitis. Injected intravenously, it gives rise to an acute septicæmia, with a blocking of the capillaries of the internal organs with microbes. If certain conditions are associated with the injections, pyæmia results, and the secondary abscesses themselves contain the streptococcus.

By successive passages through animals and cultures in the serum-bouillon media mentioned above organisms of extreme virulence may be obtained. By injections of graduated doses a condition of resistance may be established in animals, and there appears to be some ground for believing that the serum of such protected animals may itself protect other animals from streptococcic infection.

STAPHYLOCOCCUS PYOGENES AUREUS AND ALBUS

In the pus from a large proportion of abscesses, boils, pustules, and other similar suppurative foci will be found either the Staphylococcus Pyogenes Aureus or Albus—the former the more frequently. These two organisms resemble one another closely, and, indeed, might be regarded as identical, but that cultures of the former produce a golden yellow pigment wanting in the latter. In pus (Fig. 46) the staphylococci occur as irregular, usually small groups of spherical cells lying free in the fluid. The individual cells measure ·8 μ in diameter, and are easily stained by basic dyes and by Gram's method. From pus containing them the staphylococci are very readily cultivated on any of the ordinary nutrient media, and at temperatures between 14° C. to 42° C. They are facultative anærobes. On agar they form copious, thick, opaque, moist growths resembling a streak of oil paint, yellow in the case of Staphylo-

FIG. 47.

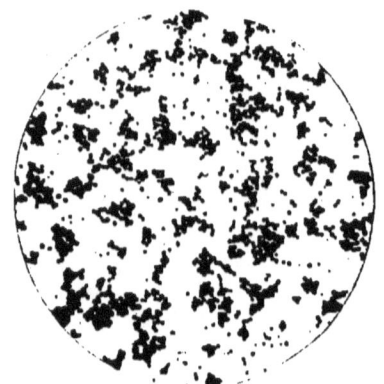

FIG. 48.

FIG. 47.—STAPHYLOCOCCUS PYOGENES AUREUS.

Gelatine Culture. 1 : 1.

FIG. 48.—STAPHYLOCOCCUS PYOGENES AUREUS.

Cover Glass Preparation, Agar Culture, Gentian-Violet, $\frac{1}{12}$ apochromat, Projection ocular 6. × 100.

coccus Pyogenes Aureus, and white in that of Staphylococcus Pyogenes Albus.

Similar growths occur on gelatine, but the medium is rather rapidly liquefied, and the character lost. A stab-culture of Pyogenes Aureus in gelatine is shown in Fig. 47, and appears as a funnel-shaped depression of liquefied gelatine rendered turbid by the suspended growth. The funnel is fairly wide, and slopes regularly and not rapidly from the top to the bottom of the culture. In time the whole tube would be liquefied. In cover-glass preparations the organism is shown in Fig. 48, and appears as a congeries of equal-sized spherical cells, showing a tendency to arrange themselves as diplococci. It is as diplococci that they generally appear when examined without staining in drop culture, while tetrads and short chains are not uncommon.

These organisms, like the streptococcus, vary considerably in their virulence, but form cultures which are much more copious and possessed of much greater vitality, so that sub-culture is easy.

The staphylococci in subcutaneous injection or inunction (even with intact skin) give rise to local suppuration but not to inflammation of the erysipelatous type. They also on intra-vascular injection produce septicæmia and pyæmia. Immunity against these infections has not been artificially produced.

MICROCOCCUS GONORRHŒÆ

THE specific purulent urethritis of Gonorrhœa was shown by Neisser to be associated with the presence of a microccccus differing from those found in ordinary pyogenic processes. By culture and inoculation the microbe was shown, chiefly by the work of Neisser and Bunn, to be the cause of gonorrhœa.

The gonococcus in gonorrhœal pus is seen (Fig. 49) to be a diplococcus occurring in groups varying from 8 or 10 to 20 or 30, and these groups are contained almost invariably within the pus cells. The individual cells are about $\cdot 8 \mu$ in diameter, and are frequently flattened or reniform, the flattened surfaces being apposed. They stain readily with basic dyes, but are differentiated from the Staphylococci of ordinary pus in that they do not stain by Gram's method. They are found in the pus of gonorrhœal ophthalmia, and also have been described as occurring in the articular fluid in cases of gonorrhœal rheumatism.

They are difficult to cultivate, grow best on media containing blood serum, and inoculations, except in the human subject, do not appear to have succeeded.

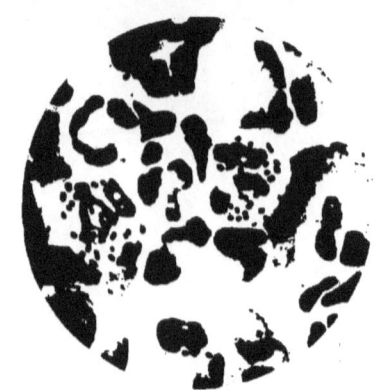

FIG. 49.

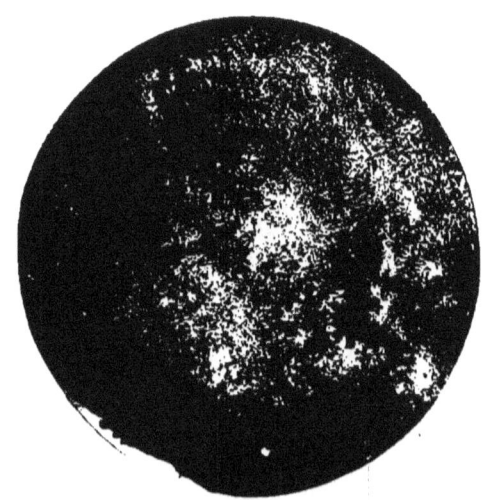

FIG. 50.

FIG. 49.—MICROCOCCUS GONORRHŒA.

Cover Glass Preparation from Urethral Pus, Löffler, $\frac{1}{12}$ apochromat, Projection ocular 6. × 1000.

FIG. 50.—B. TYPHOSUS.

Section of Mesenteric Gland, Methylene Blue, 24 mm. apochromat, Projection ocular 6. × 100.

FIG. 49.—*Micrococcus Gonokokkus*.
Cover Glass Preparation from Urethral Pus, Löffler, $\frac{1}{12}$ apochromat, Projection ocular 6., × 100.

FIG. 50.—B. Typhus.
Section of Mesenteric Gland, Methylene Blue, 24 mm. apochromat, Projection ocular 6., × 100.

BACILLUS TYPHOSUS

DISCOVERED and described by Eberth in 1880, the B. Typhosus was isolated and its causal relation to enteric fever proved by Gaffky in 1884.

In sections of the spleen, liver, or swollen mesenteric glands of patients dead from enteric fever, especially in the earlier stages of the disease, are found scattered rather large colonies of bacilli, easily seen with a low power and having frequently an arterial distribution. The tendency of the bacilli to collect into colonies in the organs is a marked feature of the B. Typhosus, and is shown in Fig. 50, which represents a section through a mesenteric gland. The two large dark patches are the colonies of bacilli. In the internal organs, as a rule, the colonies are not associated with much change in the surrounding tissues, but in the intestine inflammatory and necrotic alterations are the rule.

The tissue round the colonies of Fig. 50 is necrosed and has lost all power of staining.

If such a colony is examined with a high power it is seen to be composed of numerous short, thick, round-ended bacilli (about 2 μ by ·8 μ), which are present either as single cells or occasionally as short chains. They are generally heaped together so that individual bacilli can hardly be distinguished except at the margins (Fig. 51).

The bacilli in sections stain with methylene-blue and fuchsine, but are easily decolorised, and are therefore not easily differentiated. This organism does not stain by the method of Gram.

From such infected organs, especially the spleen and gall bladder, pure cultures may be obtained by employing any of the usual separation methods. Should gelatine plate cultures kept at 21° C. be employed, colonies develop which, at first small, rather clear, and very faintly granular, gradually, if on the surface, assume the characters seen in Fig. 52. The colony is greyish white, somewhat transparent, and iridescent. It is raised in the centre and presents a thin, spreading, festooned margin. The surface is marked by groovings and ridges which run from the centre to the periphery, but are crossed by other ridges concentrically arranged. The figure shows well the superficial markings, but exaggerates the sharpness of outline and the raised character of the

Fig. 51.—B. Tryonius.

a 4× Fig. 50, Basilin at edge of ation ocular. × 550.

Fig. 52. B. Tryonius.

Fig. 51.—B. Typhosus.

Same Preparation as Fig. 50, Bacilli at edge of Colony, ¼ apochromat, Projection ocular 6. × 550.

Fig. 52.—B. Typhosus.

Colony on Gelatine, Zeiss, 35 mm. apochromat. × 6.

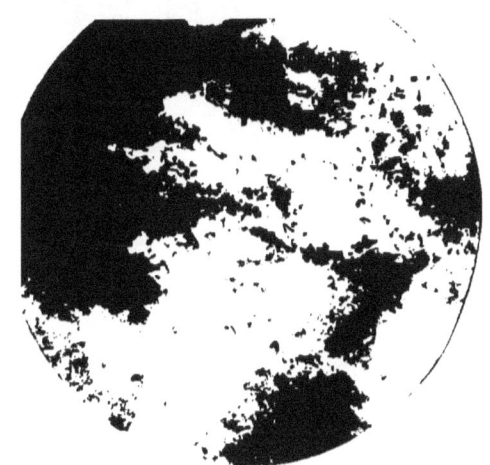

FIG. 51.

FIG. 52.

Fig. 53.

Fig. 54.

FIG. 53.—B. TYPHOSUS.

Gelatine Culture. 1 : 1.

FIG. 54.—B. TYPHOSUS.

Cover Glass Preparation, Agar Culture, 48 hours' growth, Carbol-Fuchsine, $\frac{1}{12}$ apochromat, Projection ocular 6. × 1000.

FIG. 93.—B. Typhosus

Gelatine Culture. 1:1.

FIG. 94.—B. Typhosus

Cover-Glass Preparation, Agar-Culture, 4 hours' growth.
Unstained. 1:1. Apochromat. Projection ocular 6 × 1000.

colony, which is more filmy and spreading. The gelatine is not liquefied. A streak sub-culture in gelatine shows exactly the same characters and produces a semi-transparent iridescent growth spreading over the surface of the medium with a thin filmy waved margin to be well seen on each side of the growth represented (Fig. 53).

No gas bubbles are produced when the organism is cultivated in gelatine containing glucose (compare Fig. 61, B. Coli) and milk is not coagulated. Cover glass preparations from such a culture or from similar cultures on agar show organisms like those described in the colonies (Fig. 54). They are, however, often longer than when found in tissues and chains of several apposed cells and long filamentous forms are by no means rare, especially in bouillon culture. The bacilli are, compared with B. Coli, longer, thinner, and less frequently coccoid in form.

Cultures can be obtained at temperatures ranging from 4° C.—46° C. The optimum temp. is about 37° C., and the thermal death point of the cell about 60° C.

On some media, especially potato, the organisms show a tendency to segregation of the protoplasm, so that when stained the extremities of the cell are deeply coloured and the centre is clear and apparently vacuolated. This appearance—polar staining—was supposed to be evidence of spore formation, but Buchner has shown that this is not the case. A preparation showing such polar staining with the central "clear space" is represented in Fig. 55, and the resemblance to the appearances presented by true spore formation will be seen. A culture of B. Typhosus examined in a "hanging drop" is seen to be actively motile, and is found to owe this motility to the possession of numerous long flagella. Fig. 56 shows these flagella to be many times the length of the bacillus, to be very numerous, and to pass out from all parts of the cell periphery so as to form the so-called "spider cells." The flagella are decidedly more numerous in the case of B. Typhosus than in that of B. Coli, and usually vary from about 10-18.

With pure cultures of B. Typhosus a fatal disease can be produced in many of the lower animals, though the virulence of the cultures rapidly disappears. As a rule, the disease produced is acute and toxic in its character, but recently a chronic disease resembling the enteric fever of man has been produced in rabbits by feeding them on infected food. Against the disease produced by inoculation animals may be immunised. The blood serum of

Fig. 55.—B. Typhosus.

Cover Glass Preparation, Bouillon Culture, showing Fl... staining, Carbol-Fuchsine, 1/12 apochromat, Projection ocula... × 1000.

Fig. 56.—B. Typhosus.

Cover Glass Preparation, Flagella, Nicolle and Mo... modification of Löffler's method, 1/12 apochromat, Proje... ocular 6. × 1000.

FIG. 55.—B. TYPHOSUS.

Cover Glass Preparation, Bouillon Culture, showing Polar Staining, Carbol-Fuchsine, $\frac{1}{12}$ apochromat, Projection ocular 6. × 1000.

FIG. 56.—B. TYPHOSUS.

Cover Glass Preparation, Flagella, Nicolle and Morax modification of Löffler's method, $1\frac{1}{2}$ apochromat, Projection ocular 6. × 1000.

FIG. 55.

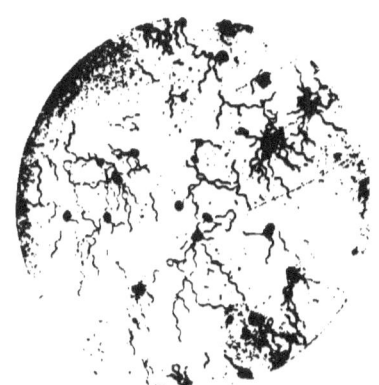

FIG. 56.

such protected animals is found, when added in small quantity to a broth culture of B. Typhosus, to produce an agglutination of the bacilli; so that they lose their motility, collect into flocculi visible to the naked eye, and are deposited in a layer at the bottom of the tube, while the supernatant fluid becomes clear. A similar action takes place if the serum of a patient suffering from typhoid fever (except in the first few days of the disease) be added to the culture. This reaction does not take place with serum from other sources (except occasionally), nor does this specific serum react thus with cultures of organisms other than the B. Typhosus. Our knowledge of these facts arose from the work of Pfeiffer on Sp. Choleræ, was extended by Gruber and Durham, and was applied by Widal to the diagnosis of enteric fever in man. The agglutinating action can be observed under the microscope. Fig. 57 represents the clumping power exerted by serum from a case of typhoid upon the bacillus. Three large masses of agglutinated bacilli are seen in the field, while the rest of the preparation shows scarcely any micro-organisms; and this clearing of the general field is almost as characteristic as the formation of clumps. A red blood cell is seen at the upper part of the field.

BACILLUS COLI

RECOGNISED as a normal inhabitant of the human intestine, and as an organism very widely spread in nature, much attention has of late been directed towards this bacillus owing to the difficulty which exists in distinguishing it from the B. Typhosus and also from the supposed pathogenic power which, under certain conditions, it possesses.

The Bacillus Coli is a short round-ended motile bacillus, ærobic, non-sporebearing, and producing no liquefaction in gelatine (Fig. 58). In morphology it is very variable, sometimes appearing as a short figure of eight bacillus, and sometimes as a distinctly cylindrical organism. In length it varies from $\cdot 8\ \mu$ to $3\ \mu$, and is about $\cdot 5\ \mu$ in thickness. Two forms are shown in Figs. 58 and 59. Compared with B. Typhosus, it is, as a rule, shorter and less bacillary, is less actively motile, and possesses fewer flagella (4–8). Like that organism, it is not stained by Gram's method. In culture in gelatine,

FIG. 57.—B. TYPHOSUS.

Agglutination by Serum from Typhoid Patient, unstained, $\frac{1}{6}$ apochromat, Projection ocular 6. × 400.

FIG. 58.—B. COLI COMMUNIS.

Cover Glass Preparation, Agar Culture, 48 hours' growth, "Bacillary" form, Carbol-Fuchsine, $\frac{1}{12}$ apochromat, Projection ocular 6. × 1000.

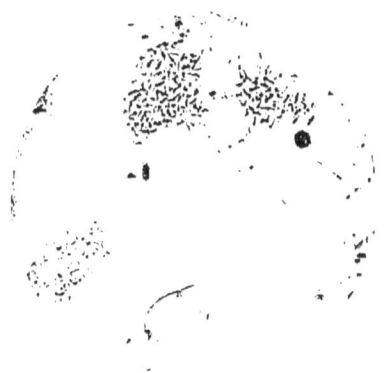

Fig. 57.

Fig. 58.

FIG. 59.

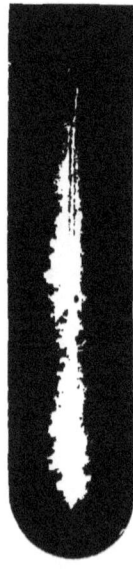 

FIG. 60.   FIG. 61.

FIG. 59.—B. COLI COMMUNIS.

Cover Glass Preparation, Agar Culture, 48 hours' growth, "Coccoid" form, Carbol-Fuchsine, $\frac{1}{12}$ apochromat, Projection ocular 6. × 1000.

FIG. 60.—B. COLI COMMUNIS.

Gelatine Culture. 1 : 1.

FIG. 61.—B. COLI COMMUNIS.

Stab Culture in Glucose-Gelatine. 1 : 1.

whether in plates or as a streak culture, the growth of B. Coli closely resembles that of B. Typhosus, but is somewhat more rapid and shows an even greater tendency to form a filmy spreading growth (Fig. 60).

B. Coli possesses the power of fermenting sugars of both the $C_6H_{12}O_6$ and $C_{12}H_{22}O_{11}$ groups, and as a consequence coagulates milk and produces gas in fluid or solid media containing these sugars. A gelatine glucose tube inoculated in stab is rapidly dislocated by the gas bubbles, and presents the appearance shown in Fig. 61. B. Typhosus, under the same conditions, produces no gas and no coagulation of milk, and the two organisms are further distinguished by the property possessed by B. Coli of producing indol in solutions of peptone and the absence of this power in the case of B. Typhosus.

DIPLOCOCCUS PNEUMONIÆ
(FRÄNKEL)

IN a large proportion of cases normal saliva contains a capsulated micro-organism, usually appearing as a diplococcus, which, when injected into the rabbit produces a rapidly fatal septicæmia. First noticed by Pasteur (1881) in the saliva, and studied by Sternberg in relation to the rabbit septicæmia produced by its injection, this micro-organism, by the work of Talamon, Fränkel, Netter, Weichselbaum, and others, was gradually connected with, and finally regarded as, the cause of croupous pneumonia in man. A normal inhabitant of the mouth, this microbe requires for the development of its pathogenic properties conditions which are at present but ill understood. The production of a typical pneumonia by its action seems to imply a considerable power of resistance in the infected animal; otherwise a septicæmia is produced. The organism possesses pyogenic properties and is found in otitis

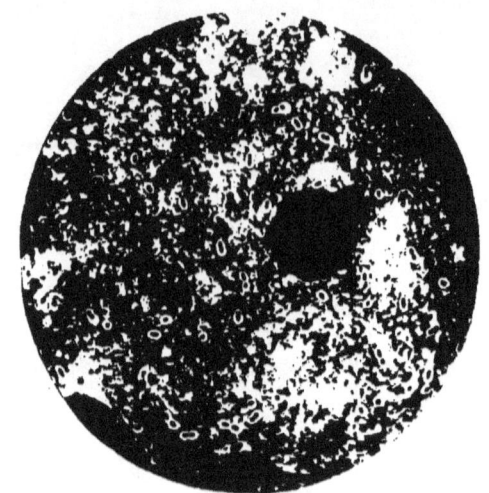

Fig. 62.

Fig. 63.

FIG. 62.—DIPLOCOCCUS PNEUMONIÆ (FRÄNKEL).

Cover Glass Preparation, Peritoneal Fluid of Inoculated Animal, Friedländer's method, $\frac{1}{12}$ apochromat, Projection ocular 6. × 1000.

FIG. 63.—DIPLOCOCCUS PNEUMONIÆ.

Agar Culture. 1 : 1.

media, cerebro-spinal meningitis, and many other diseases.

The cultures are both infective and toxic. Susceptible animals can be protected against the virulent organism, but this protection is not generally considered to extend to the toxins. Protection is afforded by the blood serum of immunised animals, or of convalescents from pneumonia (Klemperer).

As found in the saliva, in the sputum of pneumonic patients, in pneumonic lungs and in the blood, pleural and peritoneal exudations of inoculated animals the organism appears as a minute diplococcus about ·5 μ to 1 μ in diameter (Fig. 62) surrounded by a well-marked capsule. The individual cells are elongated and hastate in shape, and are placed with their broad bases in apposition when united as diplococci. This is the typical form, but variations from it are very common and many cells are simply spherical.

The organism is stained by Gram's method.

Between the temperatures 22° C. and 42° C. the microbe can be cultivated, but unless special precautions are taken the cultures are scanty, the virulence rapidly lost, and the vitality persistent for a short time only. On agar, which is much improved as a culture medium by being covered with a layer of blood, the culture (Fig. 63) occurs as a delicate growth of small, circular, discrete colonies resembling closely those of Streptococcus Pyogenes. The density of the colonies varies with the medium, but they are generally transparent and dewdrop-like with a slight thickening of the centre.

Preparations from such a culture show characteristic diplococci, but without capsule; this is lost in all forms of culture. In fluid media the microorganism occurs not only in the coccus form but also as moderately long streptococci. The preparation (Fig. 64) is from an agar-blood culture, but a few chains of 3 or 4 elements are present. The hastate shape of the individual cells is retained.

Fig. 64.—Diplococcus Pneumoniæ (Fränkel).

Cover Glass Preparation, Agar (Blood) Culture, Gram, $\frac{1}{12}$ apochromat, Projection ocular 6. × 1000.

Fig. 65.—B. Pneumoniæ (Friedländer).

Cover Glass Preparation, Peritoneal Fluid of Inoculated Animal, Gentian-Violet, $\frac{1}{12}$ apochromat, Projection ocular 6. × 1000.

Fig. 64.

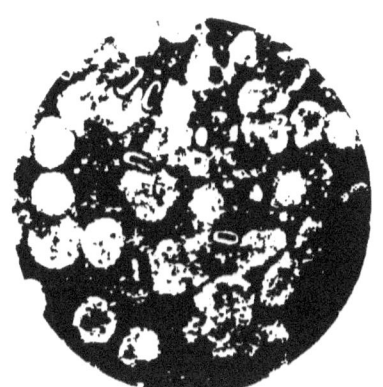

Fig. 65.

BACILLUS PNEUMONIÆ
(FRIEDLÄNDER)

IN the sputum and in the secretions from the lungs of pneumonic patients a second capsulated organism is frequently found. It was originally described and cultivated by Friedländer and thought to be the cause of croupous pneumonia. Like the preceding organism it is found in the normal mouth and associated with the formation of pus. It is larger, more bacillary, much more easily cultivated and less pathogenic than the Diplococcus Pneumoniæ, and is further distinguished by not staining by Gram's method. It gives rise when injected into the mouse to a fatal septicæmia, and Fig. 65 shows the bacillus as it occurs in the peritoneal exudation of such an infected animal; it should be compared with Fig. 62. The organism is seen to vary from an almost coccoid form (at the lower part of the figure) to a fairly long, plump, round-ended bacillus. One bacillus has become detached in the preparation and lies a little above its empty capsule. The organism grows easily and at the ordinary temperature as well as at 87° C. It forms a non-liquefactive nail-shaped growth in gelatine stab culture, a copious turbidity in bouillon, and a thick very gelatinous culture on agar. The capsule is lost in culture though the gelatinous material often found in the cultures gives rise to an appearance resembling a capsule. The preparation (Fig. 66) is from an agar culture and shows the rather broad round-ended bacillus. The length of the organism varies greatly and in all cultures many short coccoid forms occur.

FIG. 66.—B. PNEUMONIÆ.

Cover Glass Preparation, Agar Culture, Carbol-Fuchsine, $\frac{1}{12}$ apochromat, Projection ocular 6. × 1000.

FIG. 67.—B. DIPHTHERIÆ.

Section of False Membrane, Löffler, $\frac{1}{8}$ apochromat, Projection ocular 6. × 600.

Fig. 66.

Fig. 67.

BACILLUS DIPHTHERIÆ

In 1873 Klebs described amongst the organisms found in diphtheritic membranes a small bacillus which from its constancy he regarded as probably specific. This opinion was confirmed and the bacillus isolated by Löffler (1884). The observations of this investigator were confirmed and extended by Roux and Yersin by their discovery of the diphtheria poison, and the production of post-diphtheritic paralysis after inoculation of the bacillus or its toxic products. Our further knowledge, especially in regard to methods of immunisation and the antitoxic action of the blood serum of immunised animals, we owe to Fränkel and Brieger, Kitasato and Behring, and many other observers.

The bacillus is found in the superficial layer of the false membranes, and in its purest condition in the early stages of the disease. Subsequently it becomes much admixed with other organisms, especially the

Streptococcus Pyogenes, and is often crushed out by their growth.

The free surface of the false membrane is usually covered with a mixture of micro-organisms, but immediately below this layer the B. Diphtheriæ often occurs almost pure, lying in groups and small colonies in the necrosed tissues (Fig. 67). This bacillary zone is generally separated from the infiltrated deeper tissue by a necrosed but microbe-free stratum.

The bacillus is a small rod-shaped, non-motile, Gram-staining organism about the length of the tubercle bacillus (2 to 3 μ), but somewhat thicker (·7 μ). Frequently the organisms show a strong tendency to arrange themselves parallel to one another—a tendency well seen in Fig. 68, which represents a preparation made from the nasal secretion of an apparently healthy child, who, however, was the source of infection for several brothers and sisters.

The organisms, it will be observed, are to a very large extent collected within the polynuclear cells, and the photograph furnishes a good representation of the phenomenon of phagocytosis.

The bacillus, in both secretion and cultures, frequently presents a swelling and clubbing of the ex-

FIG. 68.—B. DIPHTHERIÆ.

Cover Glass Preparation, Nasal Secretion, Roux's Stain, $\frac{1}{12}$ apochromat, Projection ocular 6. × 1000.

FIGS. 69, 70.—B. DIPHTHERIÆ.

Gelatine Cultures. 1 : 1.

FIG. 68.

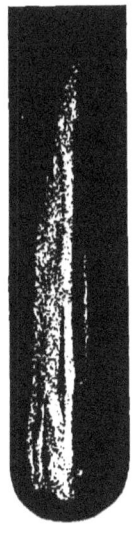

FIG. 69.

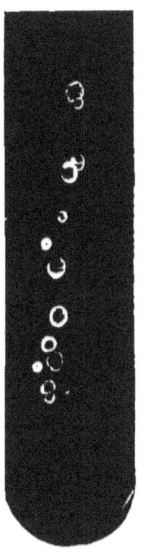

FIG. 70.

tremities which results in the production of dumb-bell and Indian-club shaped organisms. The protoplasm of the bacilli tends to segregation, so that the phenomena of "polar staining" and "banding" of the bacilli are extremely well marked, and can, together with the other features mentioned, be readily observed in the photographs.

Metachromatism is also common, and the organism is very prone to the production of so-called involution forms.

If a swab which has been rubbed over the surface of a diphtheritic membrane be then used to smear the surface of a solidified blood-serum tube, and the latter be kept at 37° C. for eighteen to twenty-four hours, it will be found covered with colonies, which in many cases consist almost exclusively of Diphtheria bacilli.

The method of bacteriological diagnosis of diphtheria by culture, which is conducted as above, depends on the extreme suitability of the medium and the temperature for the growth of the diphtheria bacillus, as compared with the other micro-organisms generally associated with the specific bacillus.

The colonies formed are discrete, circular, greyish-white growths which are thickened in the centre, have a thin and often fimbriated or fissured edge, and adhere somewhat closely to the serum. On agar and gelatine the characters of the colonies are as described on serum, but they are more transparent. Such gelatine cultures are seen in Figs. 69 and 70. The former shows a copious growth of minute colonies, and the latter isolated colonies in which the characters indicated are well marked.

Growths take place at temperatures between 20° C. and 42° C., but are very slow at the lower limit; the growth on gelatine therefore takes some time to develop. The optimum temperature is about 37° C.

A preparation made from such a serum culture as is described above is shown in Fig. 71, and in it will be noted the clubbing and banding of the bacillus, as well as the tendency to parallelism already mentioned.

A similar preparation from an agar culture is represented in Fig. 72. As a rule, the bacillus from agar, while it shows clubbed and dumb-bell forms, is usually shorter and more regular in size than the same organism grown upon blood serum, and does not show such marked banding as these latter cultures. In the photograph there are many so-called involution forms, and the preparation shows an unusual degree of irregularity.

Fig. 71—48. Diphtheria.

Cover Glass Preparation, Serum Culture, 24 hours growth
37° C. Koch — Stain, I_2 хрошромат J_2 Projection ocular 4.
× 1500.

Fig. 71—49. Diphtheria.

Cover Glass Preparation, Agar Culture, 48 hours growth
37° C. Koch — Stain, I_2 хрошромат J_2 Projection ocular 4.
× 1500.

Fig. 71.—B. Diphtheriæ.

Cover Glass Preparation, Serum Culture, 24 hours' growth (37° C.), Roux's Stain, $\frac{1}{12}$ apochromat, Projection ocular 6. × 1000.

Fig. 72.—B. Diphtheriæ.

Cover Glass Preparation, Agar Culture, 48 hours' growth (37° C.), Roux's Stain, $\frac{1}{12}$ apochromat, Projection ocular 6. × 1000.

FIG. 71.

FIG. 72.

Fig. 73.

Fig. 74.

FIG. 73.—B. DIPHTHERIÆ.

Cover Glass Preparation, Serum Culture, 72 hours' growth (37° C.), Roux's Stain, $\frac{1}{12}$ apochromat, Projection ocular 6. × 1000.

FIG. 74.—B. DIPHTHERIÆ.

Cover Glass Preparation, Gelatine Culture, 72 hours' growth at 21° C., Roux's Stain, $\frac{1}{12}$ apochromat, Projection ocular 6. × 1000.

FIG. 73.—B. DIPHTHERIAE.

Cover Glass Preparation, Serum Culture, 72 hours' growth
C., Roux's Stain, 1/4 apochromat. Projection ocular 0.
1000.

FIG. 74.—B. DIPHTHERIAE.

Cover Glass Preparation, Gelatin Culture, 72 hours' growth
at C., Roux's Stain, 1/4 apochromat. Projection ocular 0.

As was pointed out by Abbott and is seen in Fig. 73, the bacillus on blood serum grows out into extremely long filamentous forms (6 μ in length), which, together with other features, has caused this organism to be ranked by some observers with the Streptotrichæ.

Variations in the culture material cause variations in the morphology of the bacillus, but there appear to be also well-marked varieties occurring in the false membranes, which show differences both in form and virulence, and whose differential characters remain constant in successive cultures on the same nutrient materials. Among virulent bacilli are well-recognised long and short varieties.

The effect of nutrient material on morphology is shown in Fig. 74, which represents a culture on gelatine of the same virulent organism as that seen in Fig. 71. The bacillus is much shorter and stains with much greater regularity, while the "clubbing," though still present, is much less marked.

Diphtheria can be produced by the inoculation of the bacillus or its toxic products, and by various methods animals can be protected or vaccinated against either the organism or its poisons. The blood serum of these protected animals is found to be itself capable of securing the protection of other animals when injected into their tissues. The protection so secured is valid against either the virulent organism or the poisons produced by the organism. This protective or "anti-toxic" serum also possesses therapeutic properties, and secures the recovery of animals already infected. It is this serum which is so largely used in the treatment of diphtheria in man.

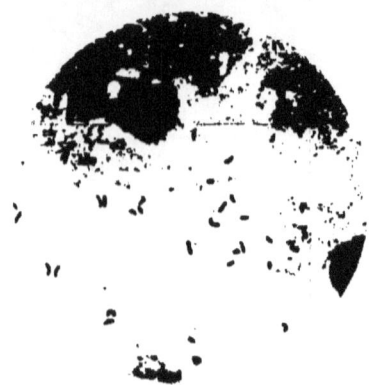

FIG. 75.

FIG. 76.

FIG. 75.—SPIRILLUM CHOLERA ASIATICA (KOCH).

Cover Glass Preparation, Rice Water Stools, Carbol-Fuchsine, $\frac{1}{12}$ apochromat, Projection ocular 6. × 1000.

FIG. 76.—SP. CHOLERA ASIATICA.

Cover Glass Preparation, Bouillon Culture, Carbol-Fuchsine, $\frac{1}{12}$ apochromat, Projection ocular 6. × 1000.

SPIRILLUM CHOLERÆ

AMONG the group of diseases caused by Spirilla the most important is Asiatic Cholera. Attributed to various miasmatic and parasitic influences it was not until 1883 that the cause of the disease was traced and identified by Koch in his study of the Egyptian epidemic of that year. His later investigation of an Indian outbreak confirmed the previous observations.

Examination of the intestinal secretions of an acute case of Asiatic Cholera and especially the mucous flakes of the "rice water" stools shows the presence, sometimes in almost pure culture, and generally in preponderating proportion, of a small comma or S shaped organism of somewhat varying thickness and of very variable curvature (Fig. 75.) The organisms are about $1\cdot5\,\mu$ in length (from $\cdot8\,\mu$ to $2\,\mu$) and $\cdot4$ to $\cdot6\,\mu$ in width. In smear preparations they are frequently arranged in groups with their long axes more or less parallel, so that the group assumes the "school of fish" appearance described by Koch. It is the simply curved comma form which is almost invariably met with in these preparations. The comma bacillus stains with fuchsine, methylene blue, or violet, but not very readily; it is not stained by the method of Gram.

From the mucous flakes found in the stools pure cultures of the organism may be obtained in various media, and a simple peptone-salt solution (1% peptone and ·5% salt) is one of the most suitable. In such a fluid medium the organism grows rapidly, renders the fluid generally turbid, and forms a thin pellicle on the surface, while the "comma" grows into a spirillum. The spirilla forms are more common in fluid than in solid media, but are by no means confined to the former, and are very variable in their length, their regularity of curvature, and the closeness of the spiral. A bouillon culture preparation, which, however, shows no long spirilla forms but some S-shaped organisms, is represented in Fig. 76, and may be compared with the preparation direct from the intestines. The preparation (Fig. 77) is made from a 48-hours-old agar culture, and shows more of the S-shaped and short spirilla forms, as well as one or two of the E-shaped organ-

Fig. 77.—Sir Cholera Asiatic.

Their Propagation. Agar Culture, 48 hours growth, with bacilli in J-shaped bunch. Projection ×875.

Fig. 78.—Sir Cholera Asiatic.

Their Propagation. Agar Culture, 24 hours growth. Sir Ho Mora. Löffler method. Flagella. Projection ×875.

FIG. 77.—SP. CHOLERA ASIATICA.

Cover Glass Preparation, Agar Culture, 48 hours' growth at 37° C., Carbol-Fuchsine, $\frac{1}{12}$ apochromat, Projection ocular 6. × 1000.

FIG. 78.—SP. CHOLERA ASIATICA.

Cover Glass Preparation, Agar Culture, 24 hours' growth at 37° C., Nicolle-Morax-Löffler method, Flagella, $\frac{1}{12}$ apochromat, Projection ocular 6. × 1000.

FIG. 77.

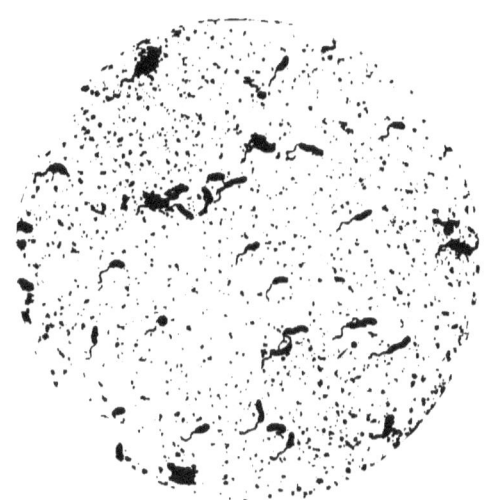

FIG. 78.

isms already referred to. Many specimens of this spirillum show, after long sub-culture, a great tendency to form very flat spirals, and assume an almost bacillary form. Klein has noted that there appears to be some material present in the intestinal secretions which gives the comma bacillus an increased power of staining, so that the flagella may be demonstrated in such preparations by simple staining methods which would not succeed with cultures in the usual media. The organism is actively motile, and owes its motility to the presence of a flagellum situated at one pole as is seen in the illustration (Fig. 78).

The varieties of the cholera spirilla differ in the number of flagella possessed by them though they are always situated at the extremity of the cell and do not usually exceed two or three. The young short comma organisms are the most actively motile, the spirilla forms being much less active, but not entirely motionless.

In some varieties (*e.g.*, Massowah) they are absent altogether.

The culture on agar forms a rapidly growing, thin, slightly iridescent film. The cultures on gelatine have, from the first discovery, been regarded as the most characteristic, and have been used to distinguish the various spirilla from one another. Unfortunately the discovery of many varieties of cholera spirilla and of many allied organisms has tended to diminish the value of this method of differentiation. The characteristic growth of the spirilla of cholera in a gelatine stab culture is shown in the two figures (Fig. 79) and (Fig. 80) representing such growth at 21° C. after two and four days respectively. Growth takes place along the whole length of the stab, but much more rapidly near the surface, and at the same time liquefaction of the gelatine takes place, so that the whole assumes the appearance of a funnel with a very long and narrow neck. The fluid appears to shrink and the funnel to be occupied by a refracting air-bubble floating on the upper part of the culture. This appearance is probably due to the clearing of the liquefied gelatine owing to the sinking of the micro-organisms to the bottom of the funnel. The collecting together and sinking of the masses of organisms is seen in both figures, but with greater clearness in the second; in this the bottom of the stab is filled with a plug of bacteria. The rate of liquefaction of the gelatine and the character of the resulting funnel of growth is very variable with the different species of Spirillum Choleræ.

FIG. 79.—SP. CHOLERA ASIATICA.

Gelatine Stab Culture, 48 hours' growth at 21° C.   1 : 1.

FIG. 80.—SP. CHOLERA ASIATICA.

Gelatine Stab Culture, 96 hours' growth at 21° C.   1 : 1.

FIG. 81.—SP. CHOLERA ASIATICA.

Gelatine Plate Culture, Colony 24 hours' growth at 21° C., 12 mm. apochromat, Projection ocular 6.   × 125

FIG. 79. FIG. 80.

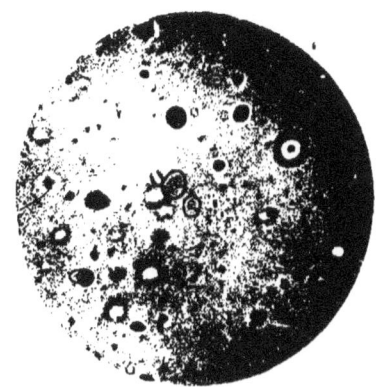

FIG. 81.

Plate cultures show signs of growth when kept at 21° C. in the first 24 hours, and considerable growth takes place in 48 hours. When the colonies are numerous the whole plate assumes a ground-glass appearance. The individual colonies are irregularly spherical, rather sharply defined, granular, clear, and highly refractile. Their appearance is well expressed by Koch's comparison which likens them to little heaps of pounded glass (Fig. 81).

At a little later stage the granular character and refractility increases, the outlines become more circular, and the edge appears very finely festooned. The surface of the colony is marked in such a way that it appears to repeat the festooning of the edge in concentric and diminishing circles. With commencing liquefaction the colony becomes surrounded by a sharp circle of fluid gelatine and the colony sinks to the bottom of the funnel-shaped hollow and appears sharply defined against the refractile zone of liquefied gelatine, while it loses its characteristic appearance.

The comma bacillus is found in large numbers in the intestinal contents in cases of cholera and occasionally penetrates the lumen of the glands and, is found between the epithelium and basement membrane; this is scarcely to be wondered at considering the intense desquamative enteritis which is set up. The organism, however, scarcely ever becomes generalised and is not found in the blood or internal organs.

That this organism is the cause of Asiatic cholera can now be scarcely doubted both on account of its constant presence in the disease and from the results of inoculation in both animals and man. It can, however, undoubtedly be present in the intestines without causing cholera, and certain little-understood conditions, *e.g.*, presence of certain microbes, intestinal irritants, disorders of digestion, &c., must exist in order that the pathogenic effects may be produced.

Animals and man may be protected against subcutaneous and intra-peritoneal inoculations of the virulent microbe, but it is doubtful if immunity can be secured against infection by the gastro-intestinal tract.

The disease appears to be toxic rather than infective.

The phenomena of agglutination similar to that which occurs with Typhoid serum and producing effects resembling those represented in Fig. 57, are readily caused by the blood serum of immunised animals, and the spirilla undergo rapid degeneration

(Pfeiffer's reaction) when injected into the peritoneal cavity of the protected animal.

No true antitoxic action has been shown to be possessed by the protective serum.

SPIRILLUM FINKLERI
SPIRILLUM AVICIDUM (METCHNIKOVII)
SPIRILLUM TYROGENUM (DENEKE)

THE three varieties of Spirilla which are illustrated in the following photographs are chiefly interesting in connection with the difficulty experienced in establishing the specific character of the Cholera Spirillum (Koch).

The Spirillum Finkleri was isolated by Finkler of Bonn. It presents morphological characters extremely like those of the true Cholera Spirillum, but is usually shorter and thicker, and less readily gives rise to spirilla forms, though it is a true spirillum. It is an extremely variable organism forming coccoid, spindle and other forms, and has in consequence been called Vibrio Proteus (Buchner). (Fig. 83).

Its culture in gelatine resembles that of Koch's

Fig. 32.—Sr. Cholera Asiatica.

Pure Plate Culture, Colony 24 hours growth, if developmental. Protection cultura 1 x 125.

Fig. 33.—Sr. Pyogenes.

Glass Preparation, Agar Culture, Gentian Violet staining, Federal monochr. 1 x 1800.

FIG. 82.—SP. CHOLERA ASIATICA.

Gelatine Plate Culture, Colony 72 hours' growth at 21° C. 12 mm. apochromat, Projection ocular 6. × 125.

FIG. 83.—SP. FINKLERI.

Cover Glass Preparation, Agar Culture, Gentian-Viole t $\frac{1}{12}$ apochromat, Projection ocular 6. × 1000.

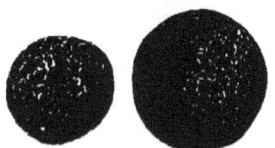

FIG. 82.

FIG. 83.

Fig. 84.

Fig. 85.

Fig. 86.

FIG. 84.—SP. FINKLERI.

Gelatine Stab Culture, 24 hours' growth at 21° C. 1 : 1.

FIG. 85.—SP. FINKLERI.

Gelatine Stab Culture, 48 hours' growth at 21° C. 1 : 1.

FIG. 86.—SP. AVICIDUM (METCHNIKOVII).

Cover Glass Preparation, Agar Culture, Watery Fuchsine, $\frac{1}{12}$ apochromat, Projection ocular 6. × 1000.

FIG. 87.

FIG. 88.   FIG. 89.   FIG. 90.

FIG. 87.—SP. AVICIDUM (METCHNIKOVII).

Cover Glass Preparation, Agar Culture, Flagella, Nicolle-Morax-Löffler, $\frac{1}{12}$ apochromat, Projection ocular 6. × 1000.

FIG. 88.—SP. AVICIDUM (METCHNIKOVII).

Gelatine Stab Culture, 48 hours' growth at 21° C. 1 : 1.

FIG. 89.—SP. AVICIDUM (METCHNIKOVII).

Gelatine Stab Culture, 96 hours' growth at 21° C. 1 : 1.

FIG. 90.—SP. TYROGENUM (DENÈKE).

Gelatine Stab Culture, 48 hours' growth at 21° C. 1 : 1.

organism, but the liquefaction proceeds so much more rapidly that by the time the characteristic growth of Sp. Choleræ is reached, the Sp. Finkleri has produced a long finger-shaped track in the gelatine, in which all trace of the original funnel has been lost. Comparison of the four photographs and of the periods of growth of the cultures they represent shows at once the nature of the differentiation (Figs. 84 and 85).

The pathogenicity of this microbe is much less than that of cholera.

The Spirillum Avicidum, or Sp. Metchnikovii, is an organism discovered by Gamaleia in the intestines of domestic fowls, and presents a very striking resemblance, both morphologically and culturally, to the organism of Koch. In preparations made from the infected animal or from pure cultures no constant differences between the two organisms can be detected, though the Sp. Metchnikovii is perhaps generally thicker and shorter. The arrangement and number of flagella is also identical (Fig. 87).

The two illustrations of the growth in gelatine might very well pass for growths of the Cholera Spirillum, though as a general rule the growth of the Sp. Avicidum is somewhat the more rapid (Figs. 88 and 89).

A further resemblance obtains between this spirillum and that of Koch in that each gives the cholera red reaction when their cultures in Peptone-salt are treated with a few drops of strong sulphuric acid. In the lower animals this organism is more virulent than the Sp. Choleræ, and subcutaneous injection in the pigeon kills the animal in 24 hours ; the spirillum is found in considerable numbers in the blood and internal organs.

The Spirillum of Deneke was isolated from decaying cheese. In culture (Fig. 85) it is very similar to the cholera spirillum, but is morphologically less like it than the two preceding organisms, and is nearly without pathogenic action.

Fig. 91.

Fig. 92.

FIG. 91.—BACILLUS PESTIS BUBONICÆ.

Cover Glass Preparation, from Liver of Rat, Carbol-Fuchsine, $\frac{1}{12}$ apochromat, Projection ocular 6. × 1000.

FIG. 92.—B. PESTIS BUBONICÆ.

Cover Glass Preparation, Agar Culture, 48 hours' growth at 37° C., Gentian-Violet, $\frac{1}{12}$ apochromat, Projection ocular 6. × 1000.

BACILLUS PESTIS BUBONICÆ

THE Bacillus Pestis Bubonicæ was discovered by Kitasato in 1894, and independently by Yersin a little later. It is found in almost pure culture in the buboes characteristic of the disease, and it is also widely spread, though not in large numbers, in the internal organs, and is found in the blood, lymph, urine, and fæces of those affected.

The bacillus is undoubtedly the cause of the disease, though the various methods by which the organism gains access to the body are not fully known; subcutaneous inoculation (wounds, &c.) is certainly one of them. Animals are affected by the disease, and many observers (Koch, &c.) think that it is primarily a disease of the rat.

Preparations made with material from the bubo, or from the spleen or liver of an infected rat (Fig. 91) show the bacillus in large numbers. It is short, thick, round-ended, and, owing to its tendency to stain more deeply at the extremities than in the centre, often presents the appearance of a diplococcus or figure-of-eight bacillus. Some examples of this polar staining can be seen in the figure. It is non-motile, occurs occasionally in short chains, is frequently capsulated, and does not stain by Gram's method.

It can be cultivated on glycerine-agar and in bouillon at 37° C. moderately well, but grows slowly and without liquefaction on gelatine at 21° C. In this medium typical thread-like colonies resembling Proteus Vulgaris are produced (Klein).

Preparation Fig. 92 is made from a young agar culture, and shows an organism resembling that from the tissues. In bouillon it assumes an almost streptococcus form, and all cultures are very prone to produce involution forms.

Animals can be protected against inoculations of the bacillus, and the blood serum of these immunised animals is both protective and therapeutic. The serum also appears to possess agglutinating properties similar to those described under B. Typhosus. Protection can be afforded by the inoculation of sterilised cultures, and, in man, vaccination by this method has been very largely employed to secure protection.

Fig. 93.

Fig. 94.

FIG. 93.—SP. OBERMEIERI.

Cover Glass Preparation from Blood of Patient, Carbol-Fuchsine, $\frac{1}{12}$ apochromat, Projection ocular 6. × 1000.

FIG. 94.—BACILLUS TETANI.

Cover Glass Preparation, Bouillon Culture, showing Spore Formation, Carbol-Fuchsine, $\frac{1}{12}$ apochromat, Projection ocular 6. × 1000.

SPIRILLUM OBERMEIERI
(RELAPSING FEVER)

In 1873 Obermeier discovered in the blood of those suffering from Relapsing Fever an actively motile, long, and very fine spirillum (Fig. 93). The organism is flexible, has finely-pointed extremities, and measures some 16 to 40 μ in length. It is stained with some difficulty by fuchsine and alkaline methylene blue. It is found only in the blood and only during the febrile attacks, and increases in numbers up to the crisis. It has not yet been cultivated.

Koch and Carter have succeeded in producing a febrile disease in monkeys by the inoculation of blood containing the spirillum, and the blood of these inoculated animals also contains the spirillum. Relapses are, however, not produced in these animals, which recover after a single attack of fever. They are not rendered immune by the inoculation.

BACILLUS TETANI

DISCOVERED in 1884 by Nicolaier in the pus from abscesses caused by the inoculation of garden earth, and isolated in pure culture by Kitasato in 1889, the specific cause of tetanus proved to be a short, fine, slowly moving anærobe, growing into long filaments and chains of shorter elements, and, under suitable conditions, producing spores. The rods stain by Gram's method, and are usually from 2-4 μ in length by about ·5 μ wide.

If the pus from a tetanus-producing wound be purified by heating to 80° C.—a temperature not destructive to tetanus spores—and then cultivated in glucose bouillon under anærobic conditions at a temperature of 37° C., a pure growth of the bacillus may be obtained. The medium becomes moderately turbid at first, but shows a strong tendency to clear and form a deposit at the bottom of the tube. If such deposit be examined after about forty-eight hours' growth, it will be found to consist largely of

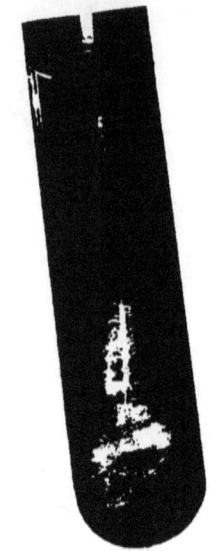

FIG. 95.

FIG. 96.

FIG. 95.—B. TETANI.

Anaërobic Culture, Glucose-Gelatine. 1 : 1.

FIG. 96.—B. ŒDEMATIS MALIGNI.

Cover Glass Preparation from Œdematous Fluid, Methylene Blue, $\frac{1}{12}$ apochromat, Projection ocular 6. × 1000.

the slowly motile vegetative cells described above, with, however, a few spore-bearing bacilli. Should the culture be older, a cover-glass preparation will show the appearance represented in Fig. 94, in which the great majority of the bacilli possess a spherical, strictly terminal spore of considerably greater diameter than the bacillus itself, giving rise to the " drum-stick " appearance so characteristic of this organism. The length of these spore-bearing bacilli is very variable, and the protoplasm, by its vacuolation and irregularity of staining, shows signs of degeneration. At the left-hand side is seen a non-spore-bearing filamentous form.

The bacillus grows between the temperatures 14° C. and 43° C. Its optimum is about 37° C. Spore formation begins at 20° C. and ceases at 42° C., and is rapid at about 37° C. Should such a culture be inoculated by means of a pipette into the depths of a glucose gelatine tube, whose surface is then melted so as to close the track of the inoculation, and whose mouth is sealed or covered with india-rubber, then, if kept at 21° C., a growth such as is represented in Fig. 95 will be obtained. The growth consists of a series of fine filaments radiating from the lower part of the inoculation track, penetrating the gelatine, and finally, but slowly, causing liquefaction, without, however, giving rise to any evolution of gas. The growth then forms a thick mass at the bottom of the tube. The same radiating filaments are seen in isolated colonies.

The pure culture, when inoculated, gives rise to tetanus. There is practically no local lesion, and the bacilli disappear rapidly from the seat of inoculation, and are not found in the blood or internal organs. The tetanus results from the toxins produced by the organism, and, while disease may be produced by the poison alone, an inoculation of the toxin-free microbe is inert and unable to produce a pathogenic effect unless assisted by chemical, traumatic, or microbic aids.

A condition of immunity may be produced in susceptible animals which is valid against either the living organism or its toxins. The blood serum of these immunised animals is capable of neutralising in "vitro" the toxin of the culture. It is also capable of preventing the pathogenic effects of inoculations of the bacillus or its toxins when the serum is injected before, at the same time, or soon after the inoculation, and it may act therapeutically after symptoms of tetanus have already shown themselves. The dose of serum required for therapeutic purposes is enormously greater than that required for prevention. Our knowledge of tetanus, the possibility of immunisation, the antitoxic and preventive action of the blood serum, is due to Kitasato and Behring, Roux, Vincent and Vaillard, Tizzoni, and others.

BACILLUS ŒDEMATIS MALIGNI

CULTIVATED soil contains not only the B. Tetani but also the B. Œdematis Maligni (Koch) or Vibrion Septique (Pasteur). Earth containing this organism when inoculated subcutaneously gives rise to a rapidly fatal disease—malignant œdema—characterised by the production of extensive subcutaneous œdema, starting from the point of inoculation and emphysema due to copious production of gas in the tissues. The same organism appears to be the cause of the occasional complication of wounds known as "acute emphysematous gangrene," "gangrène foudroyante," &c.

In the reddish fluid which fills the œdematous tissues and in the contents of the bullæ which form on the affected parts are found rather large, jointed, motile bacilli closely resembling in size and shape the B. Anthracis but distinguished from this organism by their motility, the less angular outline of the cells, and by the occurrence of chains of bacilli decidedly longer than those usually present in Splenic Fever (Fig. 96). The bacilli are from 3 μ to 6 μ in length, by ·8 μ to 1 μ in breadth. The subcutaneous fluid, as may be seen from the preparation, is remarkably rich in bacilli, which, however, are not found in the blood in any considerable numbers immediately on the death of the animal. but increase rapidly later. In pleural and peritoneal exudations the bacilli are found. From such œdematous fluids cultures may be obtained in any of the ordinary culture media (with the addition of glucose) provided the material be cultivated anærobically, as the organism is a strict anærobe. In bouillon at 37° C. rapid growth takes place with an evolution of gas and the production of a general turbidity of the medium, which subsequently clears and leaves a dense, white, flocculent mass at the bottom of the tube. From a culture on agar, preparation Fig. 97 is made, and is seen to consist of a mixture of bacilli in the vegetative form and spore-bearing bacilli. The spores are frequently in the centre of the cells, which become somewhat fusiform, but they are also often more or less terminal, and in some cases the organism closely resembles B. Tetani in appearance. It will be seen, however, that the spore is more oval and that there is always a prolongation of the protoplasm of the cell extending beyond the spore.

Contrary to what occurs in anthrax the cultures as a rule contain threads and chains of bacilli which are shorter than those found in the fluids of the inoculated animal.

The organism is not stained by Gram's method.

The growth of the bacillus is associated with the production of gas (Hydrogen and Carbonic Acid), and such a solid medium as glucose-agar is split up by gas-bubbles as is shown in photograph Fig. 98.

The gas production is often so free that the medium is quite broken up and portions of it driven up the tube.

A culture in glucose gelatine made in exactly the same way as the similar culture of tetanus (Fig. 95) is depicted in Fig. 99. A turbid sac of liquefied gelatine is formed along the line of inoculation, but stops short some little distance below the free surface of the gelatine.

The liquefaction proceeds rapidly and gas-bubbles make their appearance.

The organism is motile, and owes its motility to a large number (8–12) of peripherally arranged long flagella, which are very well seen in the figure (Fig. 100).

FIG. 97.—B. ŒDEMATIS MALIGNI.

Cover Glass Preparation, Agar Culture, showing Spore Formation, Carbol-Fuchsine, $\frac{1}{12}$ apochromat, Projection ocular 6. × 1000.

FIG. 98.—B. ŒDEMATIS MALIGNI.

Stab Anaërobic Culture, Glucose-Agar. 1 : 1.

FIG. 99.—B. ŒDEMATIS MALIGNI.

Anaërobic Culture, Glucose-Gelatine. 1 : 1.

FIG. 97.

 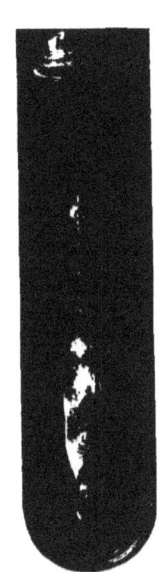

FIG. 98.   FIG. 99.

Inoculation of the pure culture produces rapid death of the animal, often within twenty-four hours. The emphysema is not so marked with the culture as after inoculation with infected soil.

It has been possible to produce immunity by the injection of filtered cultures (Roux and Chamberland), especially when the injection is made into the blood stream.

BACILLUS ANTHRACIS SYMPTOMATICI

THIS anærobic organism is the cause of "quarter evil" in cattle. The disease was at first confused with true anthrax, but the points of difference in the two affections were pointed out and the specific bacillus discovered by Bollinger and Feser in 1878.

The bacillus is a motile, fairly large organism measuring about 4 μ by 1 μ. It occurs in the tissues as single cells and as short chains (Fig. 101). It is a spore-bearing organism, but spore formation does not occur until after the death of the animal. The spores are generally central, but not unfrequently terminal, and they cause a modification in the shape of the vegetative cell similar to that which occurs in the two preceding organisms.

In culture it produces gas, and liquefies gelatine. Animals can be immunised against the action of the bacillus and vaccination of cattle is carried on abroad on a large scale.

FIG. 100.—B. OEDEMATIS MALIGNI.

*Cover Glass Preparation, Agar Culture. Fuchsin, Vu. Damascure,*¹ *apochromat, Projection ocular C. × 1000*

FIG. 101.—B. ANTHRACIS SYMPTOMATICI.

Cover Glass Preparation, Pleural Fluid, Carbol-fuchsine, J. apochromat, Projection ocular C. × 1000.

Fig. 100.—B. Œdematis Maligni.

Cover Glass Preparation, Agar Culture, Flagella, Van Ermengem, $\frac{1}{12}$ apochromat, Projection ocular 6. × 1000.

Fig. 101.—B. Anthracis Symptomatici.

Cover Glass Preparation, Pleural Fluid, Carbol-Fuchsine, $\frac{1}{12}$ apochromat, Projection ocular 6. × 1000.

Fig. 100.

Fig. 101.

ACTINOMYCOSIS

ACTINOMYCOSIS is a disease which attacks both men and cattle, producing tumours of a chronic inflammatory character, terminating in suppuration, and producing discharging abscesses of an extremely intractable character.

In the pus from these abscesses may be observed small granules of a yellowish colour, looking like portions of inspissated pus. These vary in size, attaining that of a pin's head; and when crushed beneath a cover glass are often felt to be slightly gritty. The colour varies from the usual yellow, and is sometimes greenish, brownish, or black. They were noticed by Lebert, and described by Robin in 1871. These granules are composed of the organism which is the cause of the disease. Their appearance under the microscope differs very widely in different cases, and this is often, but not always, dependent upon the source from which the pus is derived, whether Human or Bovine Actinomycosis. Fig. 102 represents what is observed when a granule from human actinomycotic pus is crushed out in a little liq. potassæ, washed in ether and alcohol, and stained by Gram's method. The nodule is seen to consist of a spherical mass of interlaced mycelial filaments, consisting of a central protoplasmic strand surrounded by a membrane. This mycelium is the essential part of the microorganism, and is alone found in the cultures (Fig. 103). The filaments grow in a radiate manner, and are long, branched, often corkscrewed, and frequently present a swelling at their extremities.

Such forms can be seen in the photographs. The mycelium stains well by Gram's method, shows no trace of division, and its protoplasm is segregated and granular. Very frequently small granules (? spores) are found in the centres of the masses which are derived by segmentation from the ends of the filaments.

Inoculated on the surface of blood serum or glycerine-agar, growth takes place at 37° C. after some days, and the inoculated granules gradually increase in size, forming a dirty white, wrinkled, raised growth. After some time, which is very variable, the growth, especially where it is most elevated and driest, becomes covered with a sulphur yellow

FIG. 102.—ACTINOMYCES HOMINIS.

Cover Glass Preparation of Pus from Actinomycotic Abscess, Gram, $\frac{1.0}{12}$ apochromat, Projection ocular 6. × 1000.

FIG. 103.—ACTINOMYCES BOVIS.

Cover Glass Preparation, Bouillon Culture, Gram, $\frac{1}{12}$ apochromat, Projection ocular 6. × 1000.

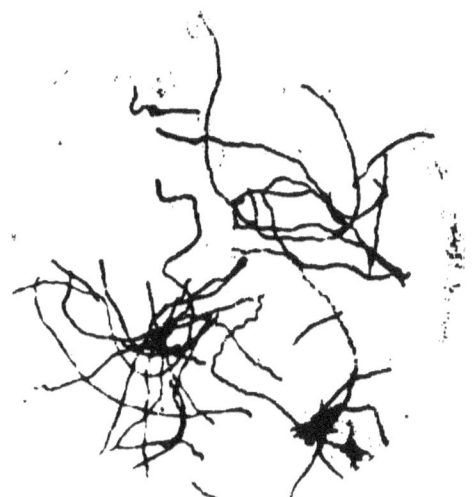

Fig. 102.

Fig. 103.

Fig. 104.

Fig. 105.

FIG. 104.—ACTINOMYCES BOVIS.

Culture on Glucose-Agar. 1 : 1.

FIG. 105.—ACTINOMYCES BOVIS.

Section of Tongue, Gram, ¼ apochromat, Projection ocular 6.
× 375.

ACTINOMYCOSIS 101

powdery deposit; and the culture assumes the appearance, except as to colour, which is exhibited by the various trichophytons. The culture material generally darkens and becomes quite brown. Such a culture on agar is shown (Fig. 104).

In granules derived from bovine actinomycosis, and in some cases of the human disease, the microscopic appearance differs much from that described above. The mycelial filaments are conspicuous by their absence, or are so reduced in length and importance as to be hardly noticed, and instead there appears a rosette of Indian-club shaped rays, which are strongly stained by Gram's method.

Fig. 105, which is photographed from a section of the tongue, shows this form of the organism. These "clubs" are formed by a swelling—whether degenerative or not is doubtful—of the sheaths covering the swollen ends of the protoplasmic filaments. The great varieties in appearance presented by the organism in different cases is due to the proportion which obtains between the ordinary mycelium and these modified club-shaped filaments, a difference further accentuated by the fact that in proportion as the mycelium increases, so does the staining by Gram's method of such clubs as are present diminish. There exists, therefore, at one end of the scale the typical bovine organism, a sphere of radiate, closely-set, clubbed filaments, strongly stained by Gram's method, and showing in the centre of the mass little or no trace of mycelium, and at the other a radiating mass of long, branching filaments, staining by Gram's method, and showing little or no trace of clubs which remain unstained. All stages intermediate between these may be found in different cases of human actinomycosis, and even in different abscesses from the same case.

Fig. 106 shows a section of pus from a case of human actinomycosis, in which the organism is intermediate in character. The dark trilobed mass is a strongly stained network of filaments lying imbedded in pus cells. At the periphery individual filaments may be distinguished, and also especially in the right lobe many clubs from which the stain has intentionally been only partially removed.

The organism has been successfully inoculated.

It usually spreads in the body by direct growth through the tissues, advancing from some centre where it has become established. It may also be distributed by the lymph and blood streams; the former is the more common. The symptoms to which it gives rise are very varied, and depend on the parts of the body affected.

Fig. 106.—Actinomyces Hominis.

Section of Hardened Pus from Actinomycotic Abscess, Gram and Eosin, ⅛ apochromat, Projection ocular 6. × 550.

Fig. 107.—Plasmodium Malariæ (Tertian).

Fully-developed Pigmented form, stained with Borax-Methylene-Blue. × 800.

Fig. 108.—Plasmodium Malariæ (Tertian).

Sporulating body, stained with Borax-Methylene-Blue. × 800.

Fig. 106.

Fig. 107.   Fig. 108.

PLASMODIUM MALARIÆ

THE parasite of malaria was discovered by Laveran in 1880, and his discovery has been confirmed by all subsequent observations. The Plasmodium Malariæ is a parasite whose habitat is the red blood corpuscle. It appears to be a protozoon, is placed by most authorities among the Sporozoa, and is closely allied to the Coccidia. It forms one of a group of similar but distinct organisms found in other vertebrata, and there are undoubted varieties, if not distinct species, of the human plasmodia, each associated with different clinical forms of the disease. The physiology and life cycle of the plasmodium regulates the phenomena of the disease, and the phases of the parasite bear a definite relation to the phases of the fever. The plasmodia have life cycles of 24, 48, and 72 hours, and give rise to quotidian, tertian, and quartan fevers respectively. In their earliest stages the parasites are found as small, pale, ill-defined, amœboid discs of protoplasm within the red blood corpuscle. These bodies increase in size, and in them appear granules of black or reddish-black pigment—melanin—derived from the hæmoglobin of the infected cell. These pigment granules, originally scattered and peripherally arranged, gradually collect into groups or radiating lines, and finally concentrate into a more or less centrally arranged mass. Around this the protoplasm of the parasite divides into regularly arranged segments, which become circular and appear as well-defined spherules possessing a vesicular nucleus and a nucleolus. These are the spores (Fig. 108), and on their maturation the red corpuscle breaks down, and the spores are set free; the melanin, which is not included in the spores, also escapes. A proportion of these free spores attach themselves to and finally enter other red cells, and the cycle begins anew. The early amœboid movements become less as the parasite matures, and cease before sporulation. After the spore enters the blood cell, as may be seen in stained specimens, the badly-staining nucleus increases in bulk together with the protoplasm, and the nucleolus comes to lie eccentrically. This eccentricity of the nucleolus, and the relative size and staining properties of the nucleus and protoplasm, give the parasite a "signet ring" (Manson) appear-

Fig. 109.

Fig. 110.

Fig. 111.

FIG. 109.—PLASMODIUM MALARIÆ (TERTIAN).

"Signet-ring" form, stained with Borax-Methylene-Blue. × 1000.

FIG. 110.—PLASMODIUM MALARIÆ (MALIGNANT TERTIAN).

"Crescent" body, stained with Borax-Methylene-Blue. × 1000.

FIG. 111.—PLASMODIUM MALARIÆ (MALIGNANT TERTIAN).

Flagellated body, stained with Carbol-Fuchsine. × 1000. (From a preparation kindly lent by Dr. Manson.)

PLASMODIUM MALARIÆ 105

ance in the corpuscle (Fig. 109). The nucleolus and nucleus gradually become less distinct, and are undistinguishable at the time of sporulation.

The above description applies to the intra-corporeal phase of the parasite, and these forms are destroyed or disappear on the administration of quinine. In the blood of patients suffering from the so-called malignant fevers, another form—the crescent body —is also found (Fig. 110). These are very regular, intra-corpuscular, crescent-shaped organisms, containing pigment which is usually, but not invariably, placed centrally; a very delicate convex line representing the red corpuscle passes from horn to horn of the crescent on its concave side. The crescent is thought by Mannaberg to be a syzigium formed by the conjugation of two plasmodia in a doubly infected corpuscle.

In blood which has been shed for some time— 10 to 30 minutes—and especially in blood which has been exposed to air and slight moisture, "flagellated bodies" are found (Fig. 111). These flagellated bodies are free, pigmented, protoplasmic masses resembling fully developed parasites from whose periphery start actively motile filaments (flagella) which break away, become free, and rapidly move about among the blood cells. The remains of the body from which the flagella are derived appear to undergo degeneration or absorption.

These flagellated bodies are derived either from the crescent bodies, or, in those varieties of the parasite which do not form crescents, from extra-corpuscular plasmodia resembling the fully developed form just prior to sporulation. The crescent becomes a sphere with central pigment, this sphere becomes agitated with movements of increasing violence, and, finally, the flagella are shot out from the periphery. Similar movements affect the flagella-producing body which is not derived from the crescent.

These flagella are regarded by Manson as the first phase of extra-corporeal life. It has been shown by Ross that the transformation and ex-flagellation take place most fully in the stomach of certain varieties of mosquito. On the analogy of observations by MacCullum on an allied parasite in crows, these flagella or flagellated spores impregnate a non-flagellated body, and the resulting organism is probably found as a pigmented cell in the stomach wall of the mosquito. The mosquito thus plays the part of secondary host, and man is infected by water or dust which has been itself infected by mosquitos. The varieties of malarial parasites at

PLASMODIUM MALARIÆ 107

present recognised are, according to Manson, from whom the following table is borrowed :

Benign { Quartan } do not form crescents.
          Tertian

Malignant { Quotidian, pigmented
            Quotidian, unpigmented } form crescents.
            Tertian

The Quartan parasite has much and relatively coarse-grained pigment. The spores are eight to ten in number, and arranged in "daisy" fashion, and there is no hypertrophy of the infected red cell.

The Benign Tertian (Fig. 107) is actively amœboid in its early stages, and causes marked hypertrophy and decolourisation of the infected corpuscle. The sporulating body is a somewhat irregular cluster of 15-20 spores, arranged round a mass of dark pigment.

The Malignant parasites are very much smaller than the Benign, and tend to assume a ringed form. Multiple infection of the red cell is common. The parasites are very numerous, but tend to pass out of the peripheral circulation into the capillaries of the internal organs and bone marrow, and the sporulating stage is rarely present in the peripheral blood. Crescents are characteristic of all the forms.

The pigmented and non-pigmented forms are

small and actively amœboid. The sporulating body consists of six to eight very small spores.

The Malignant resembles the Benign Tertian, but is smaller, does not cause hypertrophy of the red cells, and has fewer (10-12) spores, arranged in irregular heaps and rarely found in the peripheral blood.

INDEX

A

ACTINOMYCOSIS, 99
    agar culture of, 101
    "clubs" of, 101
    cultures of, 100
    inoculation of, 102
    "spores" (?) of, 100
Agar-glucose-glycerine medium for B. Tuberculosis, 36
Amphitricha, 20
Animals, how infected with anthrax, 29
Anthrax bacillus, 24
Aperture of "Iris" diaphragm in photomicrography, 5
Arthrospores, 20
Ascococci, 15
Avian tuberculosis, 38

B

BACILLI, 11, 16
    Pasteur's figure-of-eight, 16
Bacillus, The Comma, 18
Bacillus Anthracis, 24
    agar culture, 28
    animals, how infected, 29
    bouillon culture of, 27
    death point of, 27
        of spore, 27

Bacillus Anthracis, gelatine plate culture of, 28
    gelatine stab culture of, 28
    hanging drop culture, 26
    impression preparation, 28
    kidney, section of glomerulus, 31
    liver, section of, 31
    lung, section of, 30
    malignant pustule, 30
        section through, 30
    " Medusa-head " colonies of, 29
    microscopic appearance of blood of infected animal, 29
    optimum temperature of, 27
    shape of, 25
    size of, 24
    spores of, 26, 27
    spleen, section of, 31
Bacillus Anthracis Symptomatici, 98
    gas production of, 98
    immunisation against, 98
    size and motility of, 98
    spores of, 98
Bacillus Coli Communis, 64
    flagella of, 64
    fermentation, properties of, 65
    glucose-gelatine, action on, 65
    indol, production of, 65
    milk, action on, 65
    size and shape of, 64
Bacillus Diphtheriæ, 71
    agar culture of, 74
    antitoxic serum, 76
    colonies of, 73
    effects of inoculation of, 75
    gelatine culture of, 74
    involution, forms of, 73

INDEX

Bacillus Diphtheriæ, metachromatism of, 73
   morphology of, 75
   optimum temperature of, 74
   phagocytosis, 72
   polar staining and banding of, 73
   serum culture of, 74, 75
   size and shape of, 72
Bacillus Lepræ, 23, 41
   Bordone Uffreduzzi's culture of, 42
   inoculation, experiments with, 42
   size and shape of, 41
Bacillus Mallei, 46
   beading and polar staining of, 47
   cultures of, 47
   experiments with, on animals, 48
   immunisation of animals against, 48
   involution form, 47
   mallein, 48
   polymorphism of, 47
   size and shape of, 46
Bacillus Œdematis Maligni, 94
     agar culture of, 95
     bouillon culture of, 95
     effects of inoculation of, 97
     gas production of, 96
     immunisation against, 97
     motility of, 96
     shape and size of, 94, 95
Bacillus Pestis Bubonicæ, 87
    cultures of, 88
    effects of inoculation of, 87, 88
    shape of, 87
Bacillus Pneumoniæ Friedländer, 69
    gelatine-stab culture of, 70
Bacillus Smegmatis, 36, 40

Bacillus Smegmatis staining, reaction of, 40
Bacillus Syphilidis (Lustgarten), 40
Bacillus Tuberculosis, 32
 avium, 38
 branching of, 42
 " Coral Island," culture of, 37
 feeding experiments of, 32
 giant cells, 38
 inoculation experiments, 32
 Koch's method of cultivation of, 36
 optimum temperature, 37
 paths of infection of, 32
 primary cultures of, 37
 protoplasm of, 34
 saprophytic life, evidence of, 38
 secondary cultures of, 36
 segregation of, protoplasm of, 34
 size and shape of, 33
 specific staining reaction of, 33
 tuberculin, 33
 Tuberculosis, urine in, 35
Bacillus Tetani, 90
 how obtained from tetanic pus, 90
 glucose bouillon, culture in, 91, 92
 immunity discussed, 92
 inoculation, effects of, 92
 optimum temperature of, 91
 spores of, 91
Bacillus Typhosus, 59
 agglutination of, 63
 colonies in organs, 59, 60
 cultures of, 61
 death point of, 61
 effects of inoculation, 62, 63
 flagella of, 62

Bacillus Typhosus gelatine plate culture of, 60
gelatine culture of, 60
glucose gelatine, action on, 61
hanging drop culture of, 62
milk, action on, 61
optimum temperature of, 61
polar, staining of, 62
potato culture of, 61
size and shape of, 60
"spider cells," 62
"Backed" plates, their use in photomicrography, 9
Blastomycetes, 10
Bubonic plague, bacillus of, 87
Budding of B. Lepræ and B. Tuberculosis, 23

C

CELLS, lepra, 43
Centring condenser for photomicrography, 5
Cholera spirillum, 77
Classification of Schizophytes, 11
Clostridial spores, 21
Coccaceæ sub-groups, 12
Cocci, 11, 12
Coli communis bacillus, 64
flagella of, 64
gelatine culture of, 64
size and shape of, 64
Colour screens for photomicrography, 3
Condensers used for photomicrography, 2
Contrast, obtaining in negatives, 3, 4, 7, 8
"Comma" Bacillus (Koch), 18, 81
"S" and "E" shaped forms, 18
"Coral Island" culture of B. Tuberculosis, 37

"Critical Light" in photomicrography, 5
Culture tubes, how photographed, 9

D

"DAVIS" diaphragm, 6
Developer and development of negatives, 8
Diplobacillus, 16
Diplococci, 13
Diplococcus Pneumoniæ Fränkel, 66
    agar culture of, 67, 68
    size and shape of, 67, 68

E

EDWARDS's isochromatic plates, 6
Endospores, 20
    formation of, 21
Exposure in photomicrography, 6
Eye-pieces, projection, 2

F

FIGURE-OF-EIGHT bacilli (Pasteur), 16
Filaments, leptothrix, 17
    streptothrix, 17
Fission, multiplication by, 20
Flagellated organisms, 19
Focussing in photomicrography, 4
Fränkel's pneumococcus, 66
    agar culture of, 67, 68
    size and shape of, 67
Friedländer's pneumobacillus, 69
    gelatine stab culture of, 70
Fungi and fission fungi, 10

INDEX

G

GIANT cell in tubercular tissue, 38
Glanders, bacillus of, 46
    beading and polar staining of, 47
    cultures of, 47
    experiments with, on animals, 48
    immunisation of animals against, 48
    size and shape of, 46
Glycerine-glucose-agar medium for cultivation of B. Tuberculosis, 36
Gonococcus (Micrococcus Gonorrhœæ), 57
    cultivation of, 58
    size and shape of, 58
Ground glass for focussing in photomicrography, 4

H

HAND magnifier used in photomicrography, 4
Hyphomycetes, 10

I

IMPRESSION preparations of organisms, 22, 23
    preparation of colony of B. Anthracis, 28
Infection paths of B. Tuberculosis, 32
Inoculation of B. Lepræ, 42
Inoculation experiments with B. Typhosus, 62, 63
    with B. Tuberculosis, 32
Inoculation of Staphylococcus Pyogenes Aureus, 55
    of Streptococcus Pyogenes, 52, 53
Introduction, photographic, 1
"Iris" diaphragm, aperture of, in photomicrography, 5

K

KIDNEY, B. Anthracis in section of glomerulus, 31
Koch's Tuberculin, 33
Koch's method of cultivation of B. Tuberculosis, 36

L

LENSES used in photomicrography, 1
Lepræ Bacillus, 23, 41
" Lepra cells," 43
Leptothrix filaments, 17
Light "critical," 5
    filters, 3
Limelight for photomicrography, 2
Liver, section of anthrax in, 30
Löffler's Bacillus, 71
Lophotricha, 20
Lung, section of anthrax in, 30

M

MALIGNANT pustule, section through, 30
Mallei Bacillus, 46
Mallein, 48
" Medusa-Head" colonies of B. Anthracis, 29
Merismopedia, 14
Methods of multiplication of Schizophytes, 12, 20
Micrococcus GonorrhϾ, 57
    cultivation of, 58
    size and shape of, 57
Microscope and objectives used in photomicrography, 1, 2
Monotricha, 19
Morphological characters of Schizophytes, 22

INDEX

Motile organisms, 19
Multiplication by fission, 20

N

NEGATIVES, density of background discussed, 7
Nelson's quasi-achromatic condenser, 5
Nocard and Roux glycerine-glucose-agar medium for B. Tuberculosis, 36

O

OBJECTIVES and microscope used, 1, 2
Optimum temperature of B. Anthracis, 27
    B. Diphtheriæ, 74
    Streptococcus Pyogenes, 51
    B. Tetani, 91
    B. Tuberculosis, 37
    B. Typhosus, 61

P

PASTEUR, Figure-of-eight Bacilli, 16
Peritricha, 20
Pfeiffer's reaction—Spirillum Cholera, 83
Photographic Introduction, 1
Photography of culture tubes, 9
Plates, Edwards's isochromatic, 6
Plates, "backed," use of in photomicrography, 9
Plasmodium Malariæ, 103
    benign Tertian, 107
    "crescent" bodies, 105
    flagellated bodies, 105
    life cycles of, 103
    malignant parasite, 107

Plasmodium Malariæ, pigmented forms, 104
"signet" ring form, 104
spores of, 104
varieties of (Manson), 107
Projection eyepieces, 2
Protoplasm of B. Tuberculosis, 34
segregation of, 34
Pustule, malignant, section through, 30
Pyogenic organisms, 49

Q

QUARTAN parasite, malaria, 107

R

RELAPSING fever, Spirillum of, 89
Roux and Nocard, glycerine-glucose-agar medium for B. Tuberculosis, 36

S

SARCINA, 15
ventriculi, 16
Schizomycetes, 10
Schizophytes, classification of, 10, 11
morphological characters of, 22
Secondary condenser in photomicrography, 5
cultures of B. Tuberculosis, 36
Segregation of protoplasm of B. Tuberculosis, 34
Smegma Bacillus, 36, 40
staining, reaction of, 40
Specific stain of B. Tuberculosis, 33
Spirilla, 11, 17
varieties of, 18

INDEX

Spirillum avicidum (Metchnikovii), 85
Spirillum cholera asiatica, 77
    agar culture of, 79
    agglutination by serum, 82
    appearance of colonies, 81
    bouillon culture, 78
    flakes in rice stools, 77, 78
    flagella of, 79
    gelatine stab culture of, 80
        plate culture of, 81
    "school of fish" appearance, 78
    size and shape of, 77
Spirillum Deneke (Tyrogenum), 86
Spirillum Finkleri, 84
    gelatine culture, 84
Spirillum Obermeieri, 89
    size and shape of, 89
Spirochætæ, 18
Spleen, B. anthracis in section of, 31
Spore formation of B. Anthrisis, 27
Spores, clostridial, 21
Spreading inflammation, bacillus of, 50
Staphylococci, 12
Staphylococcus Pyogenes albus, 49, 54
    Pyogenes aureus, 49, 54
        agar culture of, 54, 55
        effects of inoculation and injection, 55
        hanging drop culture, 55
        gelatine culture, 55
            stab culture, 55
Streptococci, 13, 14
Streptothrix filaments, 17
Streptococcus Erisipelatosus, 51
Streptococcus Pyogenes, 49, 50
    agar culture of, 51

Streptococcus Pyogenes, bouillon culture of, 52
   effects of inoculation, 52, 53
   gelatine culture, 51
   optimum temperature of, 51

T

TUBERCULIN (Koch), 33
Tuberculosis Bacillus, 32
Tubes culture, photography of, 9
Typhosus Bacillus, 59

U

UFFREDUZZI, culture of B. Lepræ by, 42
Urine, B. Tuberculosis in, 35

V

VIBRION Septique Pasteur, 94
Vibrios, 18

W

WIDAL's reaction (B. Typhosus), 63

BOOKS ON MEDICINE
AND HOSPITALS

PUBLISHED BY

THE SCIENTIFIC PRESS, Ltd.

28 & 29 SOUTHAMPTON STREET, STRAND

LONDON, W.C.

Demy 8vo, with Two Plates, over 400 pp., cloth gilt, 12s. 6d. net.

Medical History from the Earliest Times.
By E. T. WITHINGTON, M.A., M.B. Oxon.

"One of the best attempts that has yet been made in the English language to present the reader with a concise epitome of the history of medicine, and as such we very cordially commend it."—*Glasgow Medical Journal*.

"Sets forth clearly the vicissitudes and progress of medical knowledge from the earliest times, and the gradual evolution of medical doctrines and practices up to the present century."—*Manchester Medical Chronicle*.

Demy 8vo, profusely illustrated with Coloured and other Plates and Drawings, 6s. net.

Lectures on Genito-Urinary Diseases.
By J. C. OGILVIE WILL, M.D., C.M., F.R.S.E., Consulting Surgeon to the Aberdeen Royal Infirmary, and Examiner in Surgery in the University of Aberdeen.

CONTENTS:—Urethral Fever and Catheter Fever—Treatment of Retention of Urine—Gleet and its Treatment—On Varicocele—On Hydrocele—The Treatment of Syphilis—Appendix—Prescriptions for Syphilis, etc. etc. etc.

"We have no hesitation in recommending Dr. Ogilvie Will's work to the practitioner and student of medicine."—*The Practitioner*.

Fcap. 8vo, profusely illustrated, cloth gilt, 6s.

Clinical Diagnosis: A Practical Handbook of
Chemical and Microscopical Methods. By W. G. AITCHISON ROBERTSON, M.D., F.R.C.P. (Edin.), Author of "On the Growth of Dentine," "The Digestion of Sugars," &c.

"Students and practitioners ought to welcome heartily a handbook like the one before us, which accurately and succinctly offers to one who is about to undertake a chemical or microscopical investigation just those points which he ought to know, and describes exactly the methods he ought to pursue."—*The Hospital*.

Demy 8vo, copiously illustrated, cloth boards, 3s. 6d.

Surgical Ward Work.
By ALEXANDER MILES, M.D. (Edin.), C.M., F.R.C.S.E. A Practical Manual of Clinical Instruction for Students in the Wards. Concisely, simply, and comprehensively treated. CONTENTS: Section I.—Antiseptic Surgery. Section II.—The Use of Rest in Surgery. Section III.—Bandaging. Section IV.—Surgical Instruments and Appliances.

"The book fills a distinct hiatus in surgical literature."—*Glasgow Medical Journal*.

LONDON: THE SCIENTIFIC PRESS, LIMITED
28 & 29 SOUTHAMPTON STREET, STRAND, LONDON, W.C.

18mo, profusely illustrated with original Cuts, Tables, &c.;
about 260 pp., handsomely bound in leather, 6s.

A Practical Handbook of Midwifery. By FRANCIS W. N. HAULTAIN, M.D.

A practical manual produced in a portable and convenient form for reference, and especially recommended for its compactness, conciseness and clearness.

" One of the best of its kind, and well fitted to perform the functions its author claims for it."—*Edinburgh Medical Journal.*

Royal 8vo, illustrated by a series of Original Photomicrographs and many Illustrations in the Text, cloth gilt, 7s. 6d. net.

The Menopause and its Disorders. With Chapters on Menstruation. By A. D. LEITH NAPIER, M.D., M.R.C.P., late Editor of the " British Gynæcological Journal."

" Of great practical use to the general practitioner as well as to those more specially engaged in gynæcological work."—*The Hospital.*

Fcap. 8vo, cloth gilt, 2s.

Myxœdema: and the Effect of Climate on the Disease. By A. MARIUS WILSON, M.D., B.S. L.R.C.P. (Lond.), M.R.C.S. (Eng.).

A valuable little treatise on a subject about which comparatively little has been written. The special feature of this work is its treatment of the effects of climate upon the disease.

" In this brief monograph the author outlines the features of Myxœdema as at present understood."—*New York Medical Journal.*

Second Edition. 6d.

The Schott Treatment for Chronic Heart Diseases. By RICHARD GREENE, F.R.C.P. (Edin.).

Crown 8vo, cloth gilt, 2s. 6d. net.

Dr. Mendini's Hygienic Guide to Rome.

Translated from the Italian, and edited with an additional chapter on Rome as a Health Resort. By JOHN J. EYRE, M.R.C.P., L.R.C.S. (Ireland).

" A volume at once useful and well timed."—*The Lancet.*

" May be confidently recommended, as it contains a large amount of useful information in a very accessible form not hitherto easily obtainable."
—*The Hospital.*

LONDON: THE SCIENTIFIC PRESS, LIMITED
28 & 29 SOUTHAMPTON STREET, STRAND, LONDON, W.C.

Crown 8vo, 400 pp., illustrated, cloth gilt, 5s.

Helps in Sickness and to Health: Where to go and what to do.

Being a Guide to Home Nursing, and a Handbook to Health in the Habitation, the Nursery, the Schoolroom, and the Person, with a chapter on Pleasure and Health Resorts. By Sir HENRY BURDETT, K.C.B.

" It would be difficult to find one which should be more welcome in a household than this unpretending but most useful book."—*The Times.*

" No medical or general library can be complete without such a book of ready reference."—*The Lancet.*

" This book fills a gap in popular sanitary literature by providing within the compass of one volume of very moderate size a useful collection of facts not easily found elsewhere, unless a sanitary library be at hand."—*British Medical Journal.*

Crown 8vo, profusely illustrated with 70 Drawings, cloth, 3s. 6d.

A Manual of Hygiene for Students.

By JOHN GLAISTER, M.D., D.P.H. (Camb.), Professor of Forensic Medicine and Public Health, St. Mungo's College, Glasgow.

" So much vital knowledge in so convenient a form makes this handbook very valuable. It will prove an excellent adviser to students."
Aberdeen Free Press.

Demy 8vo, with numerous Illustrations, blue buckram, gilt, 3s. 6d.

Diet in Sickness and in Health.

By Mrs. ERNEST HART, formerly Student of the Faculty of Medicine of Paris, and of the London School of Medicine for Women. With an Introduction by Sir HENRY THOMPSON, F.R.C.S., M.B. (Lond.). Fourth Thousand.

Sir Henry Thompson, in his Introduction, says:—" I do not hesitate to express my opinion that the present volume forms a handbook to the subject thus briefly set forth in these few lines, which will not only interest the dietetic student, but offer him, within its modest compass, a more complete epitome thereof than any work which has yet come under my notice."

" Mrs. Hart speaks not only with the authority derived from experience, but with the ease and freedom of an expert. We have perused this book with great pleasure, and feel sure that many will find in it much to lighten the days of those whose digestion is enfeebled."—*The Hospital.*

LONDON: THE SCIENTIFIC PRESS, LIMITED
28 & 29 SOUTHAMPTON STREET, STRAND, LONDON, W.C.

Second Edition. Crown 8vo, with 17 Plates and many Illustrations
in the Text, cloth gilt, 5s.

Infant Feeding by Artificial Means : A Scientific and Practical Treatise on the Dietetics of Infancy. By S. H. SADLER. With a new chapter on the History of Infant Feeding by Artificial Means in the Early Ages. Illustrated with Coloured and other Plates. Facsimile Autograph Letters from the late Sir ANDREW CLARK, M. PASTEUR, &c. &c.

" Mrs. Sadler's book deals with the question of the artificial feeding of infants, and contains a very useful collection of the views of the best-known English authorities upon the subject. The truly terrible ignorance displayed by mothers, especially amongst the poor, upon the subject of infant feeding is answerable for an infant mortality so great as to be appalling."—*Daily Chronicle.*

Demy 8vo, 260 pp., cloth gilt, 3s. 6d.

Art of Feeding the Invalid. By a MEDICAL PRACTITIONER and a LADY PROFESSOR OF COOKERY.

" This is a useful book. . . . To the housekeeper who has a dyspeptic, gouty, or diabetic member in her family, this book cannot fail to be of great value, and save her much anxious thought, and prevent her making serious mistakes."—*British Medical Journal.*

" Its design is excellent, and we think it has been successfully carried out, containing, as the work does, information not hitherto published in such a form. It is scarcely necessary to remark on the usefulness of information of this kind to matrons of institutions, sisters and nurses, and heads of households, and to all concerned with the care of sick and delicate people."—*The Lancet.*

" All who have to do with tending the sick will find useful guidance. The book bears evidence of skill and experience, and attention to its directions will be found better than much that goes by the name of doctoring."—*Scotsman.*

Crown 8vo, Illustrated with numerous Wood Engravings,
cloth gilt, 3s. 6d.

First Aid to the Injured and Management of the Sick. An Ambulance Handbook and Elementary Manual of Nursing. By E. J. LAWLESS, M.D., D.P.H.

" Medical officers engaged in training volunteer bearer companies, and those instructing classes in administering first aid, will find this little book invaluable."—*Pall Mall Gazette.*

LONDON: THE SCIENTIFIC PRESS, LIMITED
28 & 29 SOUTHAMPTON STREET, STRAND, LONDON, W.C.

In Four Volumes, with a Portfolio of Plans, cloth extra, bevelled. Royal 8vo, top gilt. Price as under.

Hospitals and Asylums of the World.
Their Origin, History, Construction, Administration, Management, and Legislation; with Plans of the chief Medical Institutions, accurately drawn to a uniform scale, in addition to those of all the Hospitals of London in the Jubilee Year of Queen Victoria's reign. By Sir HENRY BURDETT, K.C.B.

In Four Volumes, and a separate Portfolio containing some Hundreds of Plans.

	£	s.	d.
Price of the Book, complete	8	8	0
Vols. I. and II. (only)—Asylums and Asylum Construction	4	10	0
Vols. III. and IV.—Hospitals and Hospital Construction—with Portfolio of Plans	6	0	0
The Portfolio of Plans (20 in. by 14 in.), separately, price	3	3	0

"These magnificent volumes ... the outcome of a vast amount of laborious investigation and of many journeys in Europe, America, and the British Colonies. ..."—*National Observer.*

"The most exhaustive work on the subject extant. It is full of research historical and medical, referring to hospitals and asylums."—*Building News.*

"It would be impossible to exaggerate the value and interest of the work."—*Saturday Review.*

Third Edition, profusely Illustrated, with nearly 50 Plans, Diagrams, &c., including a Portrait of Albert Napper, Esq., the Founder of Cottage Hospitals. Crown 8vo, cloth gilt, 10s. 6d.

Cottage Hospitals, General, Fever, and Convalescent.
Their Progress, Management, and Work in Great Britain, Ireland, and the United States of America. With an Alphabetical List of every Cottage Hospital at present opened. By Sir HENRY BURDETT, K.C.B.

"A complete manual of all that relates to the founding, the cost and the management of cottage hospitals. It embodies the results of wide experience, and it demonstrates the methods by which efficiency and economy may be secured. No more complete manual could be required."—*St. James's Gazette.*

Crown 8vo, cloth, 2s. 6d.

Hospital Expenditure: The Commissariat.

LONDON: THE SCIENTIFIC PRESS, LIMITED
28 & 29 SOUTHAMPTON STREET, STRAND, LONDON, W.C.

Published Annually. Crown 8vo, over 1000 pp., cloth gilt, 5s.

Burdett's Hospitals and Charities. The Year Book of Philanthropy and Hospital Annual.

Contains a Review of the Position and Requirements, and chapters on the Management, Revenue, and Costs of the Charities. An Exhaustive Record of Hospital Work for the Year. Edited by Sir HENRY BURDETT, K.C.B.

" For fulness and for convenient arrangement of its wealth of facts this remains one of the most serviceable of all books of reference."—*Guardian*.

"This book tells us more about hospital work, medical colleges, &c., than any book we have ever seen in print. It contains more matter, more figures, more correct data that people know nothing of, than any book ever written."—*New York Medical Journal*.

Demy 8vo, profusely illustrated with Specimen Tables, Index of Classification, Forms of Tender, &c., cloth extra, 6s.

The Uniform System of Accounts, Audit, and Tenders, for Hospitals and Institutions, with Certain Checks upon Expenditure; also the Index of Classification.

Compiled by a Committee of Hospital Secretaries, and adopted by a General Meeting of the same, 18th January 1892. And certain Tender and other Forms for securing Economy. By Sir HENRY BURDETT, K.C.B.

" Removes every difficulty and sweeps away every excuse for the continuance of the muddling and obscurity which have hitherto characterised so many of these accounts."—*Statist*.

Account Books for Institutions, designed in accordance with the Uniform System of Accounts for Hospitals and Institutions.

By Sir HENRY BURDETT, K.C.B. Being a complete set of Account Books, ruled in accordance with the Uniform System adopted by the Metropolitan Hospital Sunday Fund. Designed and constructed for the convenience and assistance of Secretaries who present annual returns for participation in the Hospital Sunday Fund grants.

(*Full description of Books and Prices post free on Application.*)

LONDON: THE SCIENTIFIC PRESS, LIMITED
28 & 29 SOUTHAMPTON STREET, STRAND, LONDON, W.C.

www.ingramcontent.com/pod-product-compliance
Lightning Source LLC
Chambersburg PA
CBHW031857220426
43663CB00006B/663